CMA Monograph No. 133

FOREST-BASED CELLULOSIC MATERIALS FOR THE PAPER INDUSTRY IN INDIA

Demand-Supply Management and Pricing Policy

Tirath Gupta

OXFORD & IBH PUBLISHING CO. PVT. LTD.
New Delhi Bombay Calcutta

ISBN 81-204-0345-2

Published by Mohan Primlani for Oxford &
IBH Publishing Co. Pvt. Ltd.,
66 Janpath, New Delhi 110001 and printed at
Vishal Printers, Naveen Shahdara, Delhi 110032.

1-L8-7

Foreword

For over two decades, the Centre for Management in Agriculture (CMA) at the Indian Institute of Management, Ahmedabad (IIMA), has been engaged in research in areas like agricultural production, agricultural input supplies, cooperatives, handlooms, irrigation, fisheries, rural energy, rural development, and forestry and forest-based industries.

This study was undertaken to assess the potential demand for and supply of forest-based cellulosic raw materials for the paper and paperboards industry, and to assess the logic and feasibility of the suggested bases for pricing the cellulosic materials of forest origin and to attempt a synthesis amongst them for a long-term pricing policy.

Contrary to many earlier thoughts, it has been reasoned that the paper and paperboards industry's demand for bamboo and hardwood by A.D. 2000 could grow up to 1.7 million tonnes and 2.75 million tonnes, respectively, and to fulfil this demand it would be sufficient to harness only a portion of the estimated productive potential through conventional and non-conventional measures of resouree management.

Logic and feasibilities of a number of approaches for pricing of the cellulosic materials of forest origin have been assessed. The assessments indicated that none of the suggested bases would adequately satisfy the concerned parties. At the same time, a mix of the bases such as past and current royalty rates, general price index, market price of comparable goods could be logically used for fixing the administered prices of the natural forest produce in the short run. The most logical and viable alternative for the long run would be to raise the materials for specific manufacturing units and to price them on the basis of, but not necessarily equal to, the production costs. Some of the essential points for this purpose have been discussed.

The study further suggests that the industry may also choose to develop the alternative of entering into agreements with the farmers who have been raising tree crops. The task may be facilitated through supply of effluent water for limited irrigation, and through encouraging the farmers to raise multi-purpose tree species.

At the same time, the industry's argument for lease of land to enable them to raise the required woody materials by it needs to be assessed at the highest policy making level. It is a variant of the cost of production principle, and may also help in introducing healthy competition for objective specific rehabilitation of the waste but productive land.

It is hoped that the study will be of interest to the policy makers, administrators, industrialists, foresters, academicians and others.

Chairman
Centre for Management in Agriculture
Indian Institute of Management
Ahmedabad.

V.N. Asopa

Acknowledgements

I have had the benefit of discussions on the subject of this study with a number of our colleagues and other professionals : industrialists, administrators, forest officers, managers, economists, etc. I am thankful to them for some very valuable insights.

Shri Y. G. Jadhav, Inspector General of Forests, Government of India, Ministry of Environment and Forests; Shri R.B. Kale, formerly Chief Controller of Raw Materials with the Hindustan Paper Corporation and Ballarpur Industries; and Professor Gurdev Singh at the Indian Institute of Management, Ahmedabad (IIMA) were kind enough to make detailed comments on an earlier draft of the study. Their reviews helped in revising the manuscript. I am very grateful to them.

I am also thankful to Mrs. Geeta Chaudhuri for her editorial comments and to Shri Vinod Ahuja for proof reading and supervising the manuscript preparation.

Financial resources for the study came from the regular research grants by the Directorate of Economics and Statistics, Ministry of Agriculture, Government of India, to the Centre for Management in Agriculture (CMA) at the IIMA. I am extremely thankful to them. Similar thanks go to Professor N. R. Sheth, Director, IIMA, and to Professors Ranjit Gupta and V. N. Asopa during whose chairmanship of the CMA the work on this research project was carried out.

Centre for Management in Agriculture
Indian Institute of Management
Ahmedabad

TIRATH GUPTA

Contents

CHAPTER 1

Introduction and Objectives

India has nearly 73 million hectares, i.e., 23 per cent of its total geographical area, of forest land. But the contribution of forestry sector to the Gross Domestic Product (GDP) has been less than two per cent. The reasons for this include conservation oriented approach to the management of natural tree growth, wastage in the process of harnessing wood and non-wood products, low investments per unit of forest land resulting in a continually widening gap between expenditures and revenues from the forestry sector at least in the past, rights and privileges of the local people over a substantial portion of the forest produce, and under pricing of the raw materials of forest origin used by the manufacturing sector.

This study examines the demand-supply management and pricing policy of a set of cellulosic raw materials for the writing and printing papers, and paperboards industry in India. These raw materials can be categorised as: (i) forest-based and, (ii) non-forest-based. The former category includes bamboo, hardwoods, softwoods, grasses, and reeds. The second category includes agricultural and agro-industrial residues such as cotton stalk, wheat and rice straw, bagasse, cotton, and jute waste. Two of the goods of forest origin, i.e., bamboo and mixed woods—are most important for the paper industry and this study focusses on them.

In India, the paper industry also uses hardwood supplied by the farmers. The present study does not cover such materials even though their proportion has been rising in the recent past. Their demand and supply have been governed almost entirely by market forces.

The issues pertaining to demand-supply management and pricing of forest-based bamboo and hardwoods used as raw materials by the paper and paperboards industry are important because :

—Annual production of paper and paperboards in the country has increased from just about 0.1 million tonnes in 1950 to 1.5 million tonnes in 1985, and is expected to be around 2.7 million tonnes by the turn of the century.

—Competing demands on forest produce in general and bamboo and hardwoods in particular have been rising.

—Natural forest wealth has been dwindling which has resulted in strong perceptions of absolute scarcity of wood materials in the future.

—Conflicts between objectives of forest management like preservation and enhancement of environmental quality, and enhanced outputs of tangible goods have received unprecedented attention.

—Although those concerned seem to recognise the need for a rational pricing policy of forest products used as industrial raw materials but very little objective thinking has been done on the subject.

For the supply of bamboo and mixed woods, the state governments and paper and paperboards manufacturing companies entered into formal and informal agreements. The agreements specified the royalty rates per tonne of these goods, the extent of periodic revision in royalty rates, procedures for weighment of the materials and duration of the supply of agreed quantities of the materials, etc.

By definition, royalty can be fixed and charged for use of government property for harnessing a naturally occurring produce. In that sense, royalty is not the price of the forest produce but the two terms have been used interchangeably. Moreover, discussions with concerned people, a review of available literature, and a review of the agreements between the pulp and paper producers and the governments in a cross-section of three states of the Indian Union indicated that, till recently, very little thought was given to find a well defined basis for fixing the royalty rates. The process of royalty fixation was, however, made out to be of great consequence and considerable time was spent on negotiations whenever royalty rates were to be fixed and/or revised. The situation would have been different had the ground rules been specified and the negotiating parties could support their reasoning with facts.

A review of the past practices also indicated that the produce from natural forests was considered to be almost free and inexhaustible. In spite of long drawn negotiations, the royalty rates of bamboo and mixed woods till recently were extremely low. Moreover, the state governments have offered and continue to offer various financial concessions and facilities to forest-based industries. The logic appeared to be that industrial development of a region would offer more productive employment opportunities and would, thus, keep the local people away from destructive practices vis-a-vis the forests; and additional revenues accruing to the state by way of broadening the tax base could be ploughed back to enhance the productivity of forest areas [SICOM, 1970, p. 2]. But these have rarely been valid or practised.

Current thinking on royalty rates of forest produce is substantially different from that in the past. For instance, in the initial stages of growth of paper and paperboards industry, the provisions for periodic upward revision in royalty rates were made only in a few cases. Moreover, where such revisions were provided for, the extent of stipulated increase was marginal and even these were not always brought into effect. On the whole the royalty rates of bamboo between 1940s were revised at intervals of five to ten years, and rates thus revised were 10 to 20 per cent higher than the earlier levels. Contrary to this, the royalty rates prevalent during the early 1980s were 300 to 400 per cent higher than those during the middle of 1970s.

This change, however, does not necessarily mean that royalty rates are now being logically determined. The task is not easy, and has been made more difficult because of various reasons like, perceptions of acute scarcities of bamboo and mixed woods, rising prices of the traded portion of forest outputs in general and of timber in particular, an unprecedented desire by the forestry system and the state governments to enhance revenues per unit of the forest outputs, fast growth of forest-based industries including pulp and paper industry, enhanced competition for cellulosic materials amongst the industrial firms, and inefficiencies in the production processes, etc.

An undesirable consequence has been an environment of mistrust and confusion whereby unconscious and conscious efforts appear to have been made to evade the issues. Another consequence has been that the sellers and buyers of forest-based cellulosic materials are *indirectly* testing their bargaining powers.

An assessment of the relative strength of the parties to an economic transaction has to be a necessary consideration in pricing any product or service but, in the long run, it can not be sufficient for that purpose. Moreover, the process of testing the bargaining powers can not and should not be allowed to obstruct an objective assessment of the merits and demerits of alternative approaches to a pricing policy which must have the objective of resolving the felt and perceived conflicts amongst various categories of the users of bamboo and mixed woods, managing the demands for forest products as well as manufactured goods without necessarily resorting to direct rationing, enhancing the flow of scarce products towards most efficient uses, encouraging self-sustained development of forests and other unutilised or underutilised segments of the land resource, eliminating wide fluctuations in the supply and price of finished goods based on woody raw materials, etc.

1.1 Objective and Scope

The main objectives of this study are :

i) To assess the current and potential demand for and supply of forest-based cellulosic raw materials for the paper and paperboards industry.

ii) To assess the logic and feasibility of the various bases that have been suggested for pricing the cellulosic raw materials of forest origin.

iii) To identify a sustainable and logical basis for a long term pricing policy for forest outputs used as raw materials by the paper and paperboards industry.

Newsprint industry has not been covered in this study. It is recognised that newsprint production competes with writing and printing papers, paperboards, and allied products for cellulosic raw materials and other resources. On the contrary, demand for newsprint has been somewhat independent of or only indirectly related to the demand for paper and paperboards.

The newsprint consumption in India is mostly accounted for by sections of the urban populations which are relatively better off in terms of per capita incomes, educational levels, etc.

Moreover, expenditure on newsprint accounts for a relatively small portion of the total household expenditure. The price elasticity of demand for newspapers has been low. A good portion of the demand for newsprint in the form of magazines could even be characterised as ostentatious consumption. As a result, social, cultural, administrative, and even political implications of changes in the prices of writing and printing papers could be of much greater concern as compared to newsprint. Exclusion of newsprint may not, therefore, be considered a serious limitation to the study.

1.2 Organisation of the Study

The study is divided into five chapters. Chapter 2 summarises: (i) the growth process and the major problems faced by the paper and paperboards industry in India, (ii) practices with respect to supply and pricing of forest-based raw materials to the industry, and (iii) the current situation of confusion with regard to pricing of the forest-based raw materials. Chapter 3 deals with the current and potential demand for and supply of the cellulosic materials. Chapter 4 assesses the logic and feasibility of the suggested bases for administered prices of raw materials of forest origin, and identifies the most logical and feasible basis for a long term policy for the purpose. The last chapter contains the summary and a few important conclusions of the study.

For the benefit of those with little exposure to economic and management sciences, some basis concepts and issues in pricing goods and services have been summarised in Annexure I.

CHAPTER 2

The Setting

The chapter discusses some basic facts about and issues pertaining to :

- **—The growth and status of the paper and paperboards industry in India;**
- **—The forestry system in the country as the major producer and supplier of cellulosic raw materials used for pulping; and**
- **—The prevalent economic, socio-political and administrative environment within which policy alternatives for demand-supply management and pricing of bamboo and hardwoods used as inputs for making paper and paperboards have to be identified and assessed.**

2.1 Growth of Paper and Paperboards Industry in India

The paper and paperboards industry in India has a long and chequered history. In 1913, there were seven units manufacturing around 25,000 tonnes of paper per annum. The raw materials used were sabai grass (*Eulopsis binata*), hemp, rags, jute, imported spruce, and wood pulp. The first paper mill using bamboo as a raw material was started in 1918, and all the mills which were subsequently set up in the period of the mid-1920s to mid-1960s were bamboo-based. The Government of India encouraged this by putting import duties on some types of paper in 1925. The Bamboo Paper Industry (Protection) Act was enacted in 1932 [Ahuja, 1980, pp. 36-37].

The growth in paper making units has been rapid from 17 in the early 1950s to 136 in 1981 and to 270 in 1986. Nearly 225 of these were in the small-scale sector. Similarly, the domestic outputs of paper and paperboards increased from 1,35,000 tonnes in 1951 to 1,230,000 tonnes in 1981 and further to a little over 1,500,000 tonnes in 1986. This growth of a capital intensive industry should normally be indicative of the good health of the industry, but all does not seem to be well.

2.1.1 Growth in Industrial Capacity and Demand for Paper

The output of paper and paperboards in India rose by over 900 per cent during 1951-81, but the growth rate declined from nearly 170 per cent during 1951-61 to 114 per cent during 1961-71, and 58 per cent during 1971-81. More importantly, the growth rate in output during 1981-85 has been less than 22 per cent in spite of the fact that the country has not imported paper and paperboards since 1983.

These trends cannot be adequately explained by the continuous rise in the output levels in the base observations, particularly because an unprecedented increase in the demand for paper and paperboards could be, in fact was, anticipated for a number of reasons : rising income, emphasis on development of industry and commerce, heavy emphasis on education and adult literacy, etc. To further compound the issue (i) the share of imports in the total available paper and paperboards has declined continuously from 38 per cent in 1951 to 17 per cent in 1981; (ii) capacity utilisation in the paper industry has declined continually from 98 per cent in 1951 to around 63 per cent in 1985; (iii) growth rate in the total availability of paper and paperboards has declined continuously from 120 per cent during 1951-61 to 50 per cent during 1971-81 and further to around 13 per cent during 1981-85; and (iv) the income elasticities of demand for paper and paperboards in 1961, 1971, and 1981 were 2.7, 2.6, and 1.3 respectively.

It must, however, be noted here that the relationship between per capita income and demand cannot be uniform over time and situations. The income elasticity of demand usually declines with rise in income and aggregate consumption levels beyond a certain level. But, it is difficult to perceive that India had reached the level of a critical minimum consumption of paper and paperboards some time between 1971 and 1981. The decline in growth rate (not absolute) of NNP during the last decade should have been adequately compensated for by the increase in the proportion of adult literates from 11.5 in 1951 to 25.7 in 1981.

The average annual installed capacity per paper mill increased from 8,100 tonnes in 1951 to 19,600 tonnes in 1971 but declined continuously thereafter to reach 9,400 tonnes in 1985 (Table 2.1). This downward trend has been mainly due to the policy of encouraging small-scale pulp and paper units with the objective of enhancing the use of unconventional raw materials such as agricultural and agro-industrial residues and bagasse.[1]

The small-scale units have, however, experienced many problems. Though a detailed assessment of the nature and causes of such problems is not within the scope of this study, yet it could be said that the decision to encourage small-scale pulp and paper making units was not based on sound economic logic.[2]

The Planning Commission has been estimating the future demand for paper and paperboards in the country. A perusal of the available data and literature could not be helpful in understanding the methodology used for forecasting the demand. Moreover, estimates of future demands for these goods have not been a regular feature of the planning process, but targets for installed capacity were fixed for each plan period without exception. This would mean that aspiration or presumptions, rather than pragmatic considerations have weighed more heavily. This observation can also be supported by the fact that the Planning Commission and some other official agencies/committees have used the words *demand* and *requirements* interchangeably.

More importantly, perhaps, the planned output levels between the plan periods varied from 85 to 73 per cent of the planned capacities, and the anticipated capacity utilisation did not exhibit a consistent trend. Reasons for such variations have not been comprehensible, but they must have had significant

[1]Some have argued that the distinction between small and large-scale pulp and paper manufacturing units is not logical because there is no basic difference in the technology used.

[2]Over 40 small-scale units were reported to have closed down during the first quarter of 1986.

TABLE 2.1 : **Number of integrated pulp and paper mills in India by size groups of annual installed capacity : 1951 to 1985**

Years	*Annual Installed Capacity (tonnes)*					*Total*		*Average annual installed capacity ('000 tonnes per annum)*
	Above 20,000	*10,000 to 20,000*	*5,000 to 10,000*	*2,000 to 5,000*	*Below 2,000*	*Number of mills*	*Installed capacity ('000 tonnes)*	
1951	2	3	2	5	5	17	137.0	8.1
1961	6	4	3	8	6	27	410.0	15.2
1971	14	3	7	12	10	46	901.0	19.6
1979	19	6	23	27	31	106	1,530.0	14.4
1980	21	8	24	31	37	121	1,538.0	12.7
1981	22	9	31	33	41	136	1,657.0	12.2
1982	22	10	40	41	44	157	1,816.5	11.6
1983	23	11	41	49	51	175	1,907.0	10.9
1984	23	11	60	65	61	220	2,165.0	9.8
1985	23	15	68	80	63	249	2,350.0	9.4
1986	29	18	75	93	56	271	2,655.0	9.8

Source : 1. DCPPI, 1983e, for 1951 to 1978.
2. DCPPI, 1983a, for 1979 to 1983.
3. DCPP1, GOI, for 1984 to 1986.

bearings on costs per unit of the outputs. These, in turn, should affect the market price and demand for the final products, and the demand for inputs.

The observed outputs varied between 89 and 58 per cent of the installed capacity and, barring some exceptions, the decline has been consistent over time. Various reasons including shortages and high prices of cellulosic materials have been advanced to explain this situation,[3] but the two most important reasons appeared to be that,

—the production technology could not meet the desired standards of efficiency and efficacy, and
—the production capacity during the last at least 6 to 8 years has been substantially higher than the effective demand.

A few indications of the latter have been the rising inventories, the producers' pleas to cut down imports of newsprint by at least 1,00,000 tonnes per annum to enhance the use of home-produced creamwove paper and to remove restrictions on exports of printing and writing papers.[4] The situation of unprecedented increase in installed capacity and its underutilisation could have been avoided if the distinction between demand and requirements had been made.

However, the Planning Commission alone is, not responsible for this situation. A number of studies sponsored by government and industry have tried to forecast the demand for paper and paperboards somewhat systematically, but the demand projections have been almost invariably and significantly higher

[3]Exceptional hardships have been faced by some paper making units in the recent past. For instance, paper mills in West Bengal were getting as much as 90 per cent of their requirements of bamboo and pulpwoods from Bihar, Orissa, Madhya Pradesh, Assam, and the north-eastern region. This increased their transport costs [Dhanuka, 1982, pp. 9-10]. But, the landed cost of the materials in such cases was not known. Similarly, a paper mill in southern India got a portion of its bamboo supplies from Assam during the early 1980s and the landed cost was Rs. 1,000 per tonne, but the total quantity of the material used by the mill and its impact on the average cost of production was not known.

[4]Subsequently annual exports to the tune of 10,000 tonnes have been allowed starting from 1982.

than the observed consumption levels, particularly during 1975 and 1985.[5] Differences between the forecasted and observed consumption levels for 1980 and 1985 must have been an extremely serious problem for reasons already discussed.[6]

It must, of course, be recognised that irrespective of the ingenuity of a projection model, observed consumption of a goods would rarely be equal or be close to the estimated demand. It is more so for goods like paper and paperboards as their demand is *derived* from the demand for other goods and services. Some of the main sources for these variations in demand forecasts are differences in (i) the type of association assumed between explanatory and dependent variables: linear, log-normal, semi-log, log-log; (ii) type of data used: cross sectional vis-a-vis historical; (iii) assumed or estimated growth rates for explanatory variables: per capita income, GNP, population, income elasticity of demand, price elasticity of demand; and (iv) the number of explanatory variables included in the forecasting models. These, however, would not adequately explain the variations. The studies, therefore, lend credence to the idea that subjective thinking can play a vital role in demand projections. This can even be a major problem in the case of commodities which have multiple values such that the policy makers would be more sensitive to any perceptions of acute shortages.

There has also been a tendency to compare in absolute terms the extremely low per capita consumption of paper and paperboards in India to those of extremely high levels in the developed countries. One reason for this could be the interests of certain specific groups. Another reason could be that some international agencies have been increasingly concerned about ensuring adequate supplies and a fair distribution of paper and paper products, and consider them as essential prerequisites for

[5]This conclusion may be marginally disputed as the total availability of paper and paperboards has been treated as observed demand while a marginal and temporary shortages of these goods were reported during the latter half of 1970s.

[6]There have been considerably wide variations within the projected levels of demand for paper and paperboards for the years under reference. Such differences for the period 1975 to 2000 varied between 420 and 3,565 tonnes or 23 to 119 per cent [Gupta and Shah, 1987].

cultural and economic development. One consequence has been the lack of adequate recognition that a major portion of the demand for paper and paperboards in the economically advanced countries emanates from the proliferation of uses, and India need not or even may not emulate such consumption pattern. Another major consequence has been an almost complete lack of appreciation of the possibilities for conscious efforts at *demand management*. This should be feasible without rationing but with discrimination amongst uses and grades of paper. Textbooks, for example, may be printed on paper with lower grammes per square metre (gsm). Similarly, lower gsm writing paper could go in for notebooks used in classrooms and offices.[7] In essence, consumers are generally more interested in the surface area and hence the quality of the paper used can vary with the expected life of the use.

The substantially higher demand forecasts compared with the observed consumption have also been directly and indirectly responsible for the continuous decline in the utilisation of production capacities. More importantly, the installed capacity at the beginning of 1986 was about 1,50,000 tonnes higher than the projected capacity requirement of 23,30,000 tonnes by 1990. The low capacity utilisation has had an adverse impact on the health of the industry. Even the large-scale manufacturing units are faced with the problems of technological obsolescence with continued usage of decades old machinery and equipment.[8]

Contrary to demand management, the industry has sought and obtained policy interventions to reverse the process of slackening demand for paper and paperboards. For instance,

[7]Some have even argued that the extent of use of notebooks in the classrooms has been artificially inflated, and can be substantially reduced. Students can even be encouraged to use portable slates. Such steps would be financially and ecologically rewarding.

[8]It is estimated that at least 50 per cent of the installed capacity with the large pulp and paper mills requires modernisation not only to reverse the alarming drop in capacity utilisation but also to facilitate enhanced use of hardwoods, to substantially reduce the consumption of energy and chemicals per unit of outputs, to improve the quality of the outputs, to minimise the loss due to human error and drudgery, to ensure overall economies and reduction in production cost, etc. The cost of modernisation could be Rs. 7,500—Rs. 8,000 per annual tonne, and these could be more than recovered within a few years.

the Government has allowed duty free imports of wood chips/ wood pulp/waste paper, and has provided fiscal incentives by reducing excise duties on all varieties of paper and paperboards. Merits and demerits of such measures need to be examined, but that is not within the scope of this study.

2.2 Some Features of the Forestry System

A few generally known but basic facts are given here :

1) Nearly 75 million hectares or 23 per cent of the country's geographical area has been classified as forests, and over 95 per cent of that is owned and managed by the state forest departments.

2) Most of the forest area is managed on the principle of sustained yield and, till recently the sustained annual yield from natural forests was adequate to meet the demands from various users.

3) Numerous policy statements have emphasised that the forest area in the country should meet the community's basic requirements for amelioration of climate and maintenance (if not enhancement) of environmental quality, regulation of water flows, soil conservation, etc; and also meet the demand for industrial raw materials.

4) The vast expanse of the forests, the long gestation period of most tree crops, and synergy effects including intangible benefits of the quality and extent of vegetation could be some of the important reasons for public ownership of forest land.

5) The necessity of having a close working relationship between the forest departments and other departments of the government has been long recognised and lack of this rapport has resulted in : (i) encroachments on forest land for cultivation of seasonal crops; (ii) use of forest land for other developmental projects; (iii) inadequate tree cover bordering on near nakedness of vast tracts of land classified as forests and land owned by revenue departments and other organisations including village panchayats so much so that these have been classified as wastes; and (iv) cultivation of seasonal crops on marginal and sub-marginal lands which do not have the inherent capabilities to support a self-sustaining and dynamic system of crop cultivation and may face complete destruction in the foreseeable future.

6) The forest departments have been selling timber in bulk in the form of standing trees. *Stumpage value*, i.e., the price of standing trees is normally determined by the universally accepted principle known as *conversion return* or *residual approach*. This requires estimation of selling price of the products for which the forest produce provides basic raw materials. The residual value of the forest produce *in situ* is then derived by substracting all the relevant costs from the sale price of the processed goods. These normally include the costs on account of felling, collection/storage in forest areas, transportation to storage depots and the processing units, the cost of handling and converting the materials at the processing units, and the producers' normal profits [Gregory, 1972]. Thus, the appraised value of the material standing in the forests would be equal to its directly estimated value after processing minus the cost of converting it from a standing crop to a finished product [Duerr, 1960].

Estimation of the sale price of the final goods, and the different types of conversion and processing costs including normal profits, would be a herculean task. Norms evolved for this purpose would not be realistic due to inter-industry variations, and variations in location, age, size, technologies, etc. of the processing units within an industry. In spite of such problems, the residual value approach was used for a long time for pricing the forest products. This could mean that the forestry system was either not willing or was not able to undertake harvesting operations. The former situation would imply that the materials were adjudged to be in plentiful supply.[9]

This system of selling timber and its joint product—the mixed woods—has been discontinued in almost all parts of the country. Currently, the forest departments (FDs) and forest development corporations (FDCs) directly handle or closely supervise the felling, logging, and transportation operations up to the forest depots where the product is sold either in open auctions or by inviting tenders. Prior to announcing an auction or inviting tenders for specified lots, *upset* (base) prices of the material are usually fixed on the basis of prices obtained for

[9]Misuse or even abuse of the forest resources under the contract system which was an integral part of timber sales in the form of standing trees has been accepted by those concerned but its extent does not seem to have been documented.

comparable materials during the preceding three or more years.

On the contrary, excepting perhaps in only one state, bamboo continues to be sold to pulp and paper manufacturing companies in the form of standing crop in natural forests. The system is described in some detail in the next section.

2.3 Arrangements for Supply of Forest-based Raw Materials

For the supply of cellulosic materials for pulping, state governments and paper manufacturing companies have been entering into specific agreements and this system continues even today. Based on a perusal of such agreements in a cross-section of three states of the Indian Union during the 1930s to late 1970s and discussions with a number of people, the following features have emerged.

2.3.1 Royalty

By definition, royalty can be fixed and charged for use of government property for harnessing naturally occurring produce. In that sense royalty is not the price of the forest produce but the terms have been used interchangeably in the recent past.

Royalty rates per unit (tonne) of bamboo and hardwoods varied significantly amongst states and among the pulp and paper producing firms within the states. Some of the variations could be explained by the differences in :

- —Time, i.e., when the agreements were finalized—the royalty rates have been rising over time;[10]
- —Duration for which the agreements were stipulated to be valid;
- —Locations/accessibility and yield per hectare of the forest land leased to the paper making companies; and
- —The states' attitudes/policies towards industrial development in general and forest-based industries in particular.

Till the 1970s, the royalty for forests-based raw materials supplied to the pulp and paper industry was decided arbitrarily. Often the process of royalty fixation became a battle of wits

[10]There have been some excep tions to this general trend.

across the table between the lessors and the lessees. The lessors emphasised their desire to be fair to the lessees and argued that royalty rates should be fixed in such a way as to cover the recurring costs of administration, maintenance, and protection, and to leave a marginal profit on the capital. Likewise, the lessees also emphasised their being fair and straightforward and expected the royalty rates to enable them to carry a reasonable and proper return on the investment. Neither side could support their reasoning with relevent statistics. The process involved protracted negotiations, but the resulting loss of time and money cannot be calculated as very little of the proceedings of the meetings has been made public [Kothari, 1971, pp. 293-297].

A paper making company applied for a licence only when it was assured of sufficient supply of cellulosic materials. In this process, as soon as a FD established or announced that sufficient raw materials existed in a catchment to regularly feed a paper mill, negotiations for lease of the material began. The deal was finalised or a licence was sanctioned to a particular party with an understanding to decide the royalty at a later stage. The FDs were formally consulted in the process but the rates suggested by them were usually dubbed as *too high*.[11]

These practices can explain the cases where bamboo leases at royalty rate of rupee one per tonne along with other concessions have been made. There have also been cases where lessees were allowed to extract materials on the undertaking that they will pay the royalty fixed by the government but the rates were not fixed for many years [Sagreiya, 1971, pp. 537-38]. This also implies that excepting the costs on account of general administration and conventional silvicultural management, the state governments and the FDs considered the royalty money as *net surplus*.

Besides low royalty rates, the state governments have also been offering various other concessions to paper and paperboards manufacturing companies. These concessions include :

—Right to use other natural resources such as land and water without paying the full normal fees for such uses

[11]Some decisions to locate pulp and paper making units are reported to have been made without consulting the FDs. On the contrary, some forestry plantations have also been raised without adequate assessments of demand for the materials in their vicinity.

and without consideration of deterioration in the quality of bio-physical environment.

—Up to 50 per cent concession on the stipulated royalty rates during the first five to ten years of their operations.

—Up to 25 per cent concession on power tariffs during the initial periods.

—Sales tax holidays during the initial few years or interest free loans to pay the tax.

—Cash subsidies for locating industrial units in the less developed areas.

These measures indicate the additional reasons for low royalty rates.

2.3.1.1 Revision of Royalty Rates

An important characteristic in the agreements in the early stages of the growth of paper industry has been that only a very few of them provided for periodic upward revisions in the royalty rates. Where such revisions in the royalty rates were possible, the extent of the stipulated increases were once again nominal and ad hoc, and even these revisions were often not implemented. In the recent past, however, such revisions have been heavy and more frequent. A number of unprecedented developments have been reported which include :

—Considerable increase in royalty rates for some new companies even prior to the commissioning of their pulp and paper making units since the royalties for the existing companies had already been revised.[12]

—Upward revisions in royalty rates in some cases beyond the provisions of the agreements in force.

—Legislations empowering the state governments to revise royalty rates beyond the ceilings stipulated in the agreements in force and/or to reduce the time between such revisions.

On the whole, the royalty rates between 1940s and 1970s were revised periodically up to a maximum of 10 per cent of

[12]It is not known whether the commissioning of the production units was delayed substantially beyond the periods stipulated in the initial agreements.

the preceding level, but the differences between the royalty rates prevailing during the middle of 1970s and early 1980s have been 300 to 400 per cent.

2.3.2 Duration of Agreements

In the initial stages of the industry's growth, the agreements between the state governments and the pulp and paper making companies were stipulated to be valid for 30 to 40 years. In some cases, it was provided that the arrangements could be renewed for similar periods (not period). The agreement periods were, however, reduced to around 20 years starting in the middle of 1960s and at least one agreement made in the late 1970s was stipulated to be valid for only ten years. The main reasons for these changes appeared to be the perceptions of absolute scarcities of forest produce in the foreseeable future, and changes in the socio-politico dynamics.

2.3.3 Weighment

There have been significant differences in the methods and basis of weighing the produce which form the basis for computing royalties. They include the actual weighment at the factory gates, actual weighment in the forest depots, different running lengths per unit weight (tonne), different number of bamboos per unit weight and different volumes per unit weight. Two specific methods used for arriving at the total quantity of bamboo used in any given period without actually weighing it may be described here.

The first of these is known as the bundle-to-a-tonne ratio. A bamboo bundle is defined as 20 pieces of two metres length, i.e., 40 running metres per bundle. The foresters and the paper makers term this as a *standard* bundle. It should, however, need no emphasis to say that the so-called *standard* differed amongst bamboo species/quality and, thus, amongst forest areas. To arrive at the bundle-to-a-tonne ratio on air dry basis, a sample of bundles was usually weighed after about four weeks of harvesting the produce and the process was continued till two consecutive weighments at fortnightly intervals in a coupe were identical. Inefficiencies and malpractices often arose at this stage. Even in most objective situations, the pulp and paper producers would most likely attempt to select the lightest bundles for weighment while the functionaries would attempt to select the heaviest.[13]

The second method has been based on the concept of a *notional* tonne (NT). On this basis a specified number of running metres (RMT) of bamboo made one tonne. Once again, the number of running metres per tonne varied from state to state and even within a state depending on the thickness of the bamboo, thickness of the bamboo wall, etc. Moreover, a weighted average of different bamboo species had to be computed. This was a difficult task and provided the functionaries with temptations which were difficult to assess.

Even where bamboo was actually weighted, the bases of weights have been different : air dry weight which narmally had around 10 per cent moisture, sun dry weight which required that the material be weighed after the lapse of a certain period following harvesting, and gross weight which was the weight with full moisture.

There has been a continuous battle of words between the forestry system and the industry on issues pertaining to weights. On the other hand, almost everyone with whom we had the opportunity to discuss the matter visualised the necessity and the feasibility for a uniform basis. From the point of view of this study and for improved management of forest resources, the most important reason for a move in this direction would be the need for accurate data on outputs.

2.3.4. General Provisions of the Agreements

The terms and conditions, and rules of the agreements have varied amongst states and over time. A few illustrations would be helpful.

1) Some agreements made in the 1930s placed restrictions on the use of bamboo for purposes other than pulping, while some agreements made during the 1950s placed no such restrictions.

[13]This assumes that the functionaries on the two sides are fully and equally conscious of protecting the interests of their respective employers.

2) In the initial stages the FDs committed themselves to lease out additional forest areas if those earmarked did not generate the stipulated quantities of bamboo but such provisions are now non-existent.

3) In some cases as late as 1960s, merely the forest areas from which the lessees could harvest bamboo were listed without specifying the quantity of the materials to be obtained. More often, however, the bamboo quantities (or their range) to be availed annually by the lessees have been specified but the productive potentials of the leased areas were not systematically assessed even in such cases.

4) The agreements made during the 1930s provided that silvicultural rules and practices with regard to the leased areas could be modified by the FDs in consultation and agreement with the concerned paper making companies, but by the 1960s the former had the exclusive right to do so. Even then, the need for improved management and higher investments per unit of the natural forest land was not appreciated either by the FDs or the industry.

5) Interests chargeable by the state governments on royalty arrears and other money due from the paper making companies varied from fixed rates for the entire agreement periods to somewhat differential and probably scientifically floating bases like one to three per cent higher than the Reserve Bank's rate.

6) In some cases, the staff of the FDs have been supervising the bamboo harvesting operations in the leased areas without any financial liability on the concerned paper manufacturer. In some other cases, the manufacturers have been reimbursing the FDs for the expenditure incurred on such staff.

7) Up to 1950s, the paper making companies were not required to deposit any money as security for due performance of their contractual obligations. In the 1960s, nominal deposits in the form of interest earning securities with banking institutions were introduced and during the 1970s, substantially high deposits were required to be placed with the forest officers.

8) The stipulated procedures for solving disputes between the paper making companies and the governments have undergone a vast change from independent arbitrators and/or umpires nominated by the parties to referring the disputes exclusively

to the governments which continue to be parties to the agreements.

2.4 Environment Since Late 1970s/Early 1980s

Starting from the latter half of the 1970s in general and the current decade in particular, there have been a number of expected and unexpected changes in the half a century old relationship and arrangements between the state governments and the FDs on the one side and the pulp and paper making companies on the other.

The year 1972 could be considered as a turning point. An interim Report of the National Commission on Agriculture (NCA) observed that the rural population received a lot of subsidised supply of timber and forest materials and this resulted in a general attitude whereby the goods were given without considering the economic aspects of production. It also observed that, in the name of encouraging industrial development, forest materials have been leased out to forest industries for very long periods at very nominal or heavily subsidised royalty rates [GOI, 1972, p. 65].[14]

This has succeeded in proving that the royalty rates of forest produce were not appropriately determined, and that the prevalent system was not in the long term interests of managing the forest resources or the forest-based industries. It is now commonly recognised that adhocism in fixing the royalty rates of the cellulosic raw materials of forests origin has to be replaced. This, in turn, would necessitate replacement of the concept of royalty with price, and identification and assessment of alternative bases for pricing these materials.

The task is not easy. It has been made more difficult by the felt and perceived scarcities of basic raw materials, rising prices of the traded portion of forest outputs, a desire on the part of the FDs/FDCs/state governments to enhance revenues per unit of the output of cellulosic materials, increasing damage to the forest wealth caused by increasing biotic interference, fast growth of forest-based industries in general and pulp and paper

[14]Chapter 4 contains a discussion on the problems that have arisen by quoting the NCA out of context and from interpretations of the term subsidy to suit specific interests.

industry in particular, enhanced competition amongst the industrial firms, rising inventories of finished goods, and inefficiencies in the production process, etc.

An interplay of these variables has resulted in an environment of mistrust and confusion whereby unconscious and conscious efforts at evading the issues appear to have been made. Some examples follow :

1) A working group constituted by the Indian Paper Makers Association (IPMA) in consultation with the Ministry of Industry, and consisting of representatives from the FDCs, FDs, and paper industry has recommended that royalty of raw materials from natural forests should be fixed at the time of inception of a paper mill either on the basis of the mill's ability to pay or on the basis of cost of protection, administration and improvement/regeneration of the forest areas apportioned for the needed supplies. The relative merits and demerits of the two approaches were not assessed.

2) Some knowledgeable people have said that any increase in royalty rates of cellulosic materials will increase the cost of paper production three-fold [Bhandari, 1971, p. 66]. The reasoning appeared to be that 2.5 to 2.7 tonnes of these materials are required to produce one tonne paper. The fact, however, is that royalty of forest-based inputs has accounted for 0.3 to a maximum of 4.0 per cent of the total cost of paper and paperboards produced in India. Other things remaining the same, a 100 per cent increase in royalty rates would, therefore, enhance the cost of paper production by a maximum of 0.3 to 0.4 per cent.

3) The royalty rates of cellulosic materials must be kept low as the pulping costs have increased over time due to enhanced use of hardwoods which require higher quantities of chemicals per unit of the input [Bhandari, 1971, p. 66]. But, the landed costs of pulpwoods have generally been substantially lower than those of bamboo. The fact also remains that the proportion of mixed woods in the furnish has increased further for various reasons including relative intensity of competing demands on bamboo.

4) The supply of cellulosic materials must continue at concessional rates as these and other basic inputs including power and fuel, coal and railway freight account for over two-thirds of the cost of paper and paperboards production; and their prices

are directly or indirectly controlled by the government so that the industry can do very little by way of efficient management to check the increase in production costs [Jatia, 1983, p. 3].

The argument could carry weight if the industrial units were not making more than the 'normal' profits[15] and/or if there was rigorous price control on paper and paperboards. More importantly, production costs per unit of output can even be considerably reduced through modernisation of the old pulp and paper mills, enhanced utilisation of the installed capacities; savings on coal and electricity, chemicals, pater, etc.[16]

5) Since the central government had fixed a uniform price for 30 per cent of the white printing paper produced in the country, they can also fix the royalty rates of cellulosic raw materials for all the states.

But this is neither logical nor practical. Though forests are on the concurrent list yet management of the resources has primarily rested with the state governments, and the management practices and resulting outputs per unit of land have varied not only amongst states but also amongst forest areas within a state. Costs of production from man made and natural forests also have differed due to several reasons including variations in soil and climatic conditions. Moreover, the extent of forest resources vary significantly amongst states and there has been no definite correlation between the extent and quality of such resources in a state and the location of pulp and paper mills or other forest-based industries. More importantly, the central government had fixed the price for only 25-30 per cent of the total output of only the writing and printing paper (white paper) which accounted for only 12-15 per cent of the total paper and paperboards production.[17]

[15]Some economic and management concepts relevant to this study have been explained/summarised in Annexure 1.

[16]These issues are discussed in some detail at a later stage in this chapter.

[17]These controls were imposed through the Paper (Regulation of Production) Control Order of 1978 and were lifted in 1987.

It was reported at one stage that the GOI was thinking of levying an education cess on paper and paperboards production to create a special fund for subsidising notebooks and textbooks produced from paper purchased from the open market. The thought appears to have been shelved.

6) Long-term lease of hardwood growing areas on the lines of lease of bamboo areas should be practised [Mehta, 1981, p. 56]. But, this cannot be feasible as hardwoods are an associated product, and the paper industry cannot enjoy the right to sell the unpulpable materials from natural forest areas.

7) The industry's requirements for pulpwood should not be compared with alternative uses of the materials as each tonne of it used for proper manufacturing creates 7.7 mandays of work as against near nothing in case of its use as fuelwood [Raina, 1982, p. 16].

On the contrary, though the creation of employment opportunities is a relevant consideration yet it cannot precede the basic necessities of the people especially when a majority of them are oblivious of the social, cultural and economic significance of paper and paperboards.

8) For each tonne of paper produced, the industry paid to the government at least Rs. 3700 by way of excise duties and other taxes [Raina, 1982, p. 16]. A portion of this amount, say Rs. 500 per tonne of paper, could be passed on to the FDs to provide them with interest free financial resources to raise industrial plantations so that costs per unit of output would be low. Alternatively, excise duties and other taxes could be reduced to enhance the industry's capacity to pay higher prices for the forest produce [Mukherji, 1982a, pp. 44-45].

Some of the earlier discussion would suffice to show that this reasoning is not logical. The arguments also imply that the time honoured distinction between a tax and a cess can be forgotten.

9) The FDs are exploiting or misusing their position as monopoly suppliers of raw materials, and the only way out would be to allow the industry to produce the cellulosic materials for its own use.

The reasoning deserves serious thought at the highest level of policy making. At the same time, it may be repeated that the industry has made little contribution towards improved management or enhanced productivity of forest areas harnessed by them.

10) The royalty rates of cellulosic materials be fixed in a manner so as to equalise their landed cost for all the paper making units in the country. The argument is that paper and

paperboards are sold in a countrywide competitive market and none of the producing units should be at a disadvantage in terms of raw material costs.

However, as mentioned earlier the royalty of these inputs constituted a very small portion of the production costs which did differ considerably amongst the paper mills due to differences in their location, age, technology in use, extent of capacity utilisation, and the quantities of power and chemicals used per unit of output.[18]

11) If the paper and paperboards industry agreed to pay a fair price for cellulosic materials, management of pulpwood forests could be separated from the timber growing activity, and that would be to the advantage of the industry. The reasoning is that pulpwood has been a residue of timber management and its age and sizes have varied whereas the industry would like to use young and uniform sized materials [Seth, 1982, pp. 90-91].

This calls for reducing the crop rotation lengths but that should not necessarily require separation of the forestry system's functions for timber and pulpwood production.[19] Discussions with some experienced people indicated that the quantity and quality of timber and pulpwood produced from natural forests can be improved by management practices, such as regular thinning and pruning. More importantly, some practitioners with the industry have observed that young wood from tree species such as *Prosopis juliflora* is one of the most appropriate cellulosic materials and the cost of raising it on short rotations would not be higher than the production cost of conventional hardwoods (Chapter 3).

12) Though the royalty rates of cellulosic materials originating from government forests have been substantially enhanced yet the industry, if provided with land, cannot produce these at the current rates.

The main issue, however, is not the current royalty rates but the price at which supplies of cellulosic materials can be sus-

[18]Attempts at equalizing the landed costs of forest materials would also involve some practical problems which are discussed at a later stage.

[19]This also assumes that the forestry system can be the most efficient organisation to create and manage industrial plantations.

tained. The argument is also not relevant till such time when the industry's demand for lease of land to enable it to produce the materials for its own use is objectively assessed and finally accepted or rejected.

13) By the year 1990 or so bamboo would not be available for the paper industry and by A.D. 2000 even hardwoods would also not be supplied for that purpose as there would be more pressing demands on these products, and the government would yield to such pressures without considering the requirements of the industry. Therefore, the industry should not depend upon the forestry system for the supply of cellulosic materials for pulping [Singhal, 1982a, p. 38].

A relevant fact to the contrary, was that the state governments continued to be keen to attract the manufacturing units including pulp and paper mills with the objectives of encouraging developmental activities in remote rural areas and broadening the tax base. And for this purpose they extended cash subsidies and various other incentives. It has even been alleged that such policies are responsible for location of private and public sector paper mills without adequately ascertaining the availability of raw material within the financially or managerially feasible distances. The result in some cases has been the diversion of available raw materials from the existing manufacturing units to the new ones not only within the states but also across administrative boundaries of the states [IPMA, 1982, p. 13; Joshi, 1982a, p. 42].[20]

14) Paper mills have not been accepting (i) hardwoods less than two inches in thickness; and (ii) woods from the tree species such as Ain (*Hardwickia binata*), Mahua (*Madhuka indica*), Tendu (*Diospyros melanxylon*), and Dhaora (*A. latifolia*). These rejected materials constitute about 40 per cent of the total production of hardwoods and should therefore be put to use [Sowani, 1982a, pp. 38-89]. There are at least four considerations for resolving this issue.

First, the thinner materials could be classified as unpulpable

[20]The magnitude of such cases could not be ascertained but even limited examples of such practices would be contrary to the economic principles and also the desirability of meeting the commitments to the existing paper mills (not necessarily at the on-going royalty rates) for at least such periods till the recovery of investments.

due to higher costs for handling and processing than for the thicker lots. Data for such cost differences have, however, not been available. Even the basic data like differences in debarking costs of thinner and thicker materials have not been documented.

Second, the thinner materials may contain immature cellulose and the quality of pulp produced from them may not be comparable with that produced from the thicker materials. The need and importance of assessing the quality of cellulose in the thinner materials have been recognised [Sowani, 1982b, pp. 56-57] and considerable research work on pulping qualities of hardwoods has been done, but the issue continues to be debated without adequate statistical support.

Third, the hardwoods considered unpulpable by some paper mills were readily accepted by others, and it was argued that this anomaly had been caused by almost total monopoly enjoyed by some producers over the forest resources in their localities, and their unwillingness to explore appropriate technologies [Sowani, 1982a, pp. 38-39]. Once again, the discussions on technological issues have been based on opinions rather than on facts so much so that even technical feasibility of portable chippers continues to be debated.

Fourth, such a situation could not have continued to hold good had the materials been really scarce, or if the first two aspects of the issue were adequately assessed in the context of inter-state, intra-state, and intra-industry differences in the relevant variables.

15) Wood as a source of energy has attracted considerable attention and could gain further importance in the future. Since fuelwood and pulpwood are substitutes, every state must determine the proportion of total resources that could be needed for pulping, and the extent to which fuelwood requirements should be met on priority basis. Similarly, the FDs in each state should take a balanced view of the total bamboo resources and should set apart a portion of the output for pulping after meeting the local demands [GOI, 1978b, p. 10].

A number of questions pertaining to administrative, managerial, and economic feasibilities of the recommendation can be raised. An important question would be whether the solutions can be left entirely to each state government/FD. The answer has to be in the negative as (i) paper and paperboard are nationally and internationally traded products, and (ii) regional specialisation in economic activities can make positive contribution to the pace of economic development of the country. It has even been argued that with proper planning of forest operations and rational location of manufacturing units, large economies can be affected and the prices of cellulosic materials can be reduced [Thapar, 1974, pp. 51-53].[21]

16) The price of cellulosic materials should be a non-issue because if the industry pays more, the price of the finished product would also go up. Hence, it would not matter even if the price of these materials is increased by another Rs. 500 per tonne [Singhal, 1982a, pp. 38-39]. The issue, however, is not so simple. This thinking, indicates the desire of some people that, just as paper manufacturing companies have been making huge 'net' profits, the forestry system should also enhance its 'net' revenues without adequate efforts at improved management of the resources at its command.

17) Since the industry is dependent on the forestry sector for the supply of cellulosic raw materials, the former should provide the latter with an incentive to invest more and produce more by paying the maximum possible price for the forest produce. Higher prices for the forest produce would generate the financial resources needed for maintenance and improved management of natural forests as well as to raise manmade plantations [Mukherji, 1982, p. 4].

However, the FDs have to compete with other government departments for budgetary allocations, and there has not been and may not be a direct relationship between the revenues generated by and investment funds made available to them. It has also been noted that the gap between the revenues generated by the forestry sector and the funds made available to it (including normal expenditure) has been increasing over time.[22]

[21]This appeared more feasible since the pulp and paper industry in India offered a good example of non-integrative and ad hoc growth as manufacturing units have been located without adequately assessing the access to raw materials and economies of scale.

[22]One state government decided to create an industrial plantations fund out of the additional revenues to be received from the proposed royalty rates of bamboo and hardwoods. But, details regarding the (i) manner for computing the additional revenues, (ii) the proportion of additional reve-

18) The royalty once fixed should be operative for a period of at least five years of actual use of the material to automatically condone the delays in starting production. The logic is that frequent revisions of royalty rates cause uncertainties with respect to raw material supplies, pulp production costs, and profitability of the mills. The relationship between the periodicity of revision in royalty rates/price of forest produce and uncertainties in supplies could be difficult to appreciate. In fact, there have been cases of prolonged negotiations for revisions of royalty rates without interruptions in supplies of the materials.

19) Revision of royalty rates every three to five years, as has been the case in the recent past, have been difficult to absorb, and there should be a gradual rise in price/royalty of forest produce through annual revisions [IPMA, 1983; Joshi, 1982a, p. 43].

This may be logical but, if a mutually acceptable basis is not clearly laid out, the process of annual revisions of royalty (not price) would be time consuming.

20) Periodic revisions in royalty rates/prices of forest products must appropriately account for :

—The royalty fixed at the inception of a pulp and paper mill;
—Increase, if any, in the cost of protection, administration, and improvement/regeneration of forest areas apportioned for supplies of pulpable materials during the intervening period;
—Changes in the all-India consumer price index for agricultural labour during the intervening period; and
—Changes in the cost of other inputs such as power and electricity, chemicals, etc.

The list of these variables can be unmanageably long. The main questions, however, pertained to (i) the availability of the relevant data, and (ii) the weights to be assigned to the chosen variables. But, very little thought has been given to these aspects.

The discussion done, so far would suffice to show that the

nues to be placed in the proposed fund, and (iii) agency to control and operate the proposed fund, etc. have not been available.

sellers and buyers of cellulosic materials are *indirectly* testing their relative bargaining power. Relative bargaining strength of the parties to an economic transaction has to be a necessary consideration in pricing any economic good or service, but it cannot be sufficient for a long term and sustainable pricing policy. More importantly, decisions must be based on objective assessments of the relevant variables but that is not easy under the circumstances. Yet an attempt to assess the relative merits and demerits of alternative approaches for determining and periodic revision of the *price* and *not royalty*, of forest-based cellulosic materials has been made in Chapter 4.

Chapter 3

Current and Potential Demand for and Supply of Forest-based Cellulosic Materials

The main objective of this chapter is to assess the paper and paperboard industry's demand and supply of cellulosic raw materials of forest origin by the turn of the century. This would be one of the basic considerations for a long-term pricing policy, and should also serve as a basis for evaluating the alternatives for enhancing their supplies or managing their demand.

Similarly, it may be recognised that perceptions of acute shortages of bamboo and mixed woods in the foreseeable future, if valid, would be a function of (i) the *effective demand* for paper and paperboards, (ii) availability and relative price of non-forest-based cellulosic materials, (iii) administrative and managerial feasibility for enhanced use of other forest-based materials such as grasses and reeds, (iv) changes in the policy pertaining to imports of pulpable materials and ready pulp, (v) policies regarding exports and imports of paper and paperboards, (vi) extent and success of efforts at enhanced biomass production per unit of forest land and cultivated and uncultivated wastelands, and (vii) technological developments to utilise the outputs from more tree and grass species for pulping, etc.

3.1 Demand for Paper and Paperboards by A.D. 2000

The demand estimates for paper and paperboards in India by A.D. 2000 varied from 3.25 to 4.40 million tonnes as against the 1985 output of 1.5 million tonnes. A review of literature, however, showed that the demand has been substantially overtimated since 1951. On the basis of this review, qualitative observations on the markets for paper and paperboards, and discussions with knowledgeable people, it has been estimated that the demand in the foreseeable future may rise by 3.5 per cent a year [Gupta and Shah, 1987]. This means that *without* any imports and *without* any attempts at demand management, the paper and paperboard markets in the country should be in a state of near equilibrium if the production could gradually rise to 2.5 million tonnes by A.D. 2000.[23]

This estimate was commensurate with the targetted output of paper and paperboards during the Seventh Five Year Plan [GOI, 1985b, p. 188]. However, in the context of the installed capacity at nearly 3.0 million tonnes, production capacity in the pipe line, recent policies to encourage consumption, and possibility of some exports; it could be advisable to work with a growth rate in demand at 4.0 per cent per annum. This entailed a production of 2.7 million tonnes by the turn of the century.

3.2 Some Basic Considerations and Issues in Supply of Cellulosic Materials

A number of cellulosic materials such as bamboo, softwood, hardwood, grasses, reeds, agricultural residues, agro-industrial residues, cotton and jute waste, and used paper and paperboards have been and continue to be used. A few general issues may be taken up at this stage to enable us to concentrate on the prospects of supply of two of the forest-based raw materials : bamboo and mixed woods which are the focus of this study.

1) All cellulosic materials are not homogenous. Their properties in terms of quality and the cost of processing are considerably different. Similarly, paper and paperboards are not homogenous products. On the whole, however, the cellulosic materials are substitutable and the degree of substitution has been considerably enhanced by technological developments. In actual practice, the level of substitution would be a function of a number of variables, and chief among them would be

[23]The mix of outputs is not relevant to this discussion.

the feasibility of changes in and/or adaptability of the pulping and paper making technologies in use, technologies to be used by new or modernised units, and institutional interventions including pricing policy to encourage or discourage the use of specific raw materials.[24]

2) Uniformity in the stock of any cellulosic raw material is desirable from financial and management angles. This, however, should not necessarily mean that the quantity and quality of forest-based materials can be enhanced only through man-made plantations of single species. In the context of other concerns such as ecological and environmental costs of manufacturing activities, and the problems experienced by the FDCs during the last decade or so, it may appear desirable to allocate a major portion of the limited technical and financial resources towards better management of natural forests. This is not to argue that man-made plantations should be halted. The point is that costs and returns of raising industrial plantations must be assessed in the context of costs and returns of enhancing the outputs from natural forests.

3) Amongst the most important forest products used for pulping, bamboo has become relatively more scarce in the last one decade. This situation can be eased through changes in the raw material mix.[25] Considerable work towards identifying pulpable tree species have been carried out by the Forest Research Institute (FRI), some research organisations, and also by a few paper mills. The proportion of hardwoods used by large paper mills varied from 40 to 80 per cent [IPMA, 1982a, p. 14; Raina, 1982a, p. 3]. There could, therefore, be ample scope for enhanced utilisation of these materials, and the change may be accelerated by differential pricing of pulp woods.[26]

Besides gradually reducing the proportion of bamboo in the

[24]Pulp produced from some cellulosic raw materials associated with seasonal and annual crops has been exempted from excise duties. The industry has argued that this benefit be extended to all forest-based raw materials excepting bamboo and hardwoods.

[25]The overall proportion of bamboo used by the industry stands around 60 per cent of the cellulosic materials, but it varies from 50 to 100 per cent amongst the pulp and paper making units.

[26]The thrust at this stage is not to support or refute the extent to which the royalty rates have been or are proposed to be revised.

raw material mix, the pulp and paper industry can also help itself and the society by increasing pulp yield per unit of the materials. This has been found feasible. For instance, it has been reported that prophylactic treatment of stored bamboo by spraying two per cent sodium pentachlorophenate or two per cent boric acid would cost Rs. 6 to 8 per tonne of bamboo, and would more than pay for itself through enhanced pulp yield, higher strength of the pulp, and lower use of chemicals for bleaching.[27]

4) One may argue that ready pulp can be imported at a comparatively lower cost to meet the shortage of cellulosic raw materials.[28] The current policy favours this to some extent. However, this may be unrealistic in the long run as free exchange of consumer goods such as paper and paperboards between countries and regions may not be relied upon because of a number of reasons including disparities in stages of economic development and consumption patterns, and differences in political and economic ideologies. The issue needs an indepth examination and is not within the purview of this study.

5) Besides bamboo and mixed woods, forests also generate grasses and reeds. It is argued by some that these materials cannot assume a place of importance because of their seasonal availability and low productivity per unit of land [Raina, 1982, p. 7]. The contrary should have been true if the industry was actually faced with acute shortages of pulpable materials. Improved management practices including harvesting, baling, and infrastructural development would have been adopted to augment their supplies at lower per unit costs.

Moreover, systematic cultivation of grass species such as *Eulopsis binata* is expected to be a complementary measure to soil conservation in hilly areas. With this objective, the paper and paperboard industry may be encouraged to use more of such materials. Examples of the use of grasses and reeds by a few paper mills are available, and the prospects of harnessing such materials could improve if systematic efforts are made to

[27]Practitioners doubted the effectiveness of the suggested treatment unless bamboo was protected against rain, and that was not considered to be cost effective.

[28]Pulp imports on large scale can also result in considerable savings in environmental costs.

grow them and such other crops as mesta along with tree species, and the industry is also encouraged to harness them on a sustained basis.[29]

The approach of this study, however, has been to err on the lower side of potential raw material supplies. It may, therefore, be assumed that this source would make only a marginal contribution to the total supply of cellulosic materials by the turn of the century.

6) Coniferous materials, commonly known as softwoods, are the most appropriate and form the base for pulping and paper making in many leading countries. The estimated area covered by the softwood tree species in India ranges from four to six per cent of the total forest area [IEMR, 1980, p. 144; Tiwari, 1982a, p. 11]. These forests are confined to the high Himalayan region in Jammu and Kashmir, Himachal Pradesh and Uttar Pradesh; and these will have to be conserved for a variety of reasons including protection of the fragile mountain eco-system. Wherever such forests are exploitable, harvesting and transportation costs are generally prohibitive and the limited quantities which are extracted have a substantially higher value as sawn timber compared with any other use.

The NCA had envisaged large scale plantations of tropical pines in the plains with special emphasis on the Bastar Dome, but the activity did not take off in spite of heavy investments. Thus, softwoods may not be considered a profitable source of cellulosic materials in India. This should, once again, strengthen our basic approach to err on the lower side of raw material supplies by A.D. 2000.

7) Varying levels of concessions in excise duties have been made on production of paper grade pulp with 50 per cent and above of what have been termed as non-conventional raw materials such as straw, bagasse, jute and hemp waste, and gunny cuttings.[30]

Considering the area placed under cultivation of rice, sugar-

[29]Adequate planning would also be needed to ensure that every mill does not count everybody's fodder as its raw material as seems to have happened with the boom of small paper mills.

[30]The excise concession is not applicable to some of the farm residues such as cotton stalk, castor stems, and banana stems. A discussion on such issues is not within the scope of this study.

cane and jute such cellulosic materials have been presumed to be of considerable importance. India is stated to be the second largest producer of wheat and rice straw with the estimated annual output of 150 million tonnes. A substantial number of small scale paper mills established during the last eight to 10 years have been based on these materials. In actual practice, however, there have been many expected and unexpected problems in enhanced industrial uses of these materials.

For instance, at least one paper mill in the organised sector located in a sugarcane belt and expected to utilise bagasse for pulping had to initially acquire the material from another state covering a distance of 500 to 600 km; and hence had to at least temporarily switch over to the conventional raw materials soon after the initial operations. Similarly, there have been considerable problems in the availability of straw for the pulp and paper mills in south Gujarat where many of the units have been working substantially below their rated capacities and have, thus, been financially unviable [DCPPI, 1983c, p. 17].

This is not to say that these materials cannot contribute to pulp and paper making. Detailed studies on the opportunity costs of these materials in the alternative uses; and problems of logistics including their handling, transportation, and storage costs are needed.

For instance, rice and wheat straw are important fodder materials. The country has 20 per cent of the world's animal population but it is not adequately fed. Similarly, bagasse is used as fuel in the sugar manufacturing process. Besides its opportunity cost in terms of equivalent calorific value of coal, the following aspects must also be looked into : (i) the availability of bagasses per tonne of cane crushed by sugar mills with varying capacities, technologies and age; (ii) financial, economic, and managerial aspects of collection and storage; (iii) financial and economic aspects of changes in sugar production technology including replacing the existing boilers; (iv) logistics and opportunity cost of ensuring adequate supplies of coal to the sugar factories; and (v) prospects of vertical integration of sugar production and pulp production processes.

Meanwhile, it can be hypothesised that vertical integration of sugar production and pulp production processes would considerably enhance the financial and managerial feasibilities of using bagasse for paper and paperboards production. However, in the context of the study's approach to err on the lower side of the raw material supplies, it is assumed that agricultural and agro-industrial residues would make only a marginal contribution to the total cellulosic raw material supply.

8) In spite of the facts that: (i) large quantity of used paper in India is reused for packaging, particularly in retail trade; and (ii) the pulp and paper making units in the country, with one exception, are not equipped to handle printed waste paper for want of de-inking facilities; an estimated 20 per cent of the total production of paper and paperboards is currently available for recycling. The world average was estimated at around 40 per cent [DCPPI, 1983b, p. 7].

Recycling waste paper offers at least two basic advantages: lower energy consumption, and negligible pollution costs in the pulping process. Financial costs of waste paper mainly comprise collection and sorting, and the societal costs on these counts would be substantially lower. In recognition of such advantages, waste paper has been classified as a non-conventional raw material for the pulp and paper industry in India; and excise concessions on pulp produced from this input have been available since 1983-84. Duty free imports of waste paper under OGL have also been permitted.

It can, therefore, be safely estimated that at least 30 to 35 per cent of the total paper and paperboards production in the country can be available for recycling. This is considered feasible because the main sources of waste paper are printing presses, offices, educational institutions, etc. and the material can be gathered at low cost. The use of this material can be further enhanced with policy interventions to facilitate installation of deinking systems.[31]

Waste paper yields 0.75 per cent pulp. It can, thus, be estimated that this source would support 0.70 million tonnes of the total expected output of paper and paperboards in India by the turn of the century.

[31]That could be particularly desirable on account of protecting the environment.

3.3 Demand for Forest-based Cellulosic Materials by A.D. 2000

The preceding discussion would suffice to show that out of the total estimated production of paper and paperboards in India to meet the demand at the turn of the century, a maximum of 1.5 to 1.7 million tonnes would be based on conventional raw materials of forest origin i.e. bamboo and hardwoods. In the context of the expressed desire to err on the positive side of demand for these raw materials, we may work with the higher figure of 1.7 million tonnes.[32]

There is almost a unanimity that the mix of bamboo and hardwood in the total of these two materials would, at worst, be 40 : 60. Inputs of bamboo and hardwood with the chemical pulp process are taken at 2.5 and 2.7 tonnes respectively, per tonne of output. This means that by the year A.D. 2000 the pulp and paper industry's demand for bamboo and mixed woods should grow to a maximum of 1.7 million and 2.75 million tonnes, respectively. The ratio of bamboo and hardwoods may even be 30:70 in which case the industry's demand for these materials would be 1.3 million and 3.2 million tonnes, respectively. But, this may not be feasible or even necessary. We would, therefore, work with the 40 : 60 ratio and the estimated demand at 1.7 million tonnes and 2.75 million tonnes of the two materials.

3.4 Supply Prospects of Forest-based Cellulosic Materials by A.D. 2000

The estimated demand for bamboo and hardwood and the size of land in the forestry system in India could lead to the conclusion that cellulosic raw materials of forest origin should not be critically scarce. As stated earlier, the forests in India encompass 74.8 million hectares, i.e. nearly 23 per cent of the geographical area of the country. Out of this, over 45 million hectares were being exploited and another 16 million hectares were considered potentially exploitable. Thus, less than 14 million

[32]The input levels would be significantly lower with other pulping processes such as mechanical or chemo-mechanical. Changes in these directions are expected to take place in the foreseeable future but have not been considered in this study. This, once again, is in line with the stated desire to work with higher side of the raw material requirements.

hectares or 20 per cent of the total forest area was in the unexploitable category.

The challenge as well as the opportunity lies in the fact that in spite of plenty of sunshine, long growing season, and adequate rainfall for tree growth, the mean annual increment of natural forests in India was estimated at 0.5 cu m per hectare[33] as compared to 2.6 cu m in Asia, 2.5 cu m in Europe, 2.3 cu m in USA, 1.9 cu m in USSR, and 2.1 cu m for the world. Similarly, the growing stock in India's forests was estimated at 25 cu m per hectare against the world average of 210 cu m per hectare [Tiwari, 1982b, p. 10]. Poor stocking of a sizeable portion of the forest land and ruthless exploitation with inefficient technology were some of the main reasons for the low growth and low output situation. For instance, at least 16-20 million hectares or 20-30 per cent of the area classified as *managed forests* was either barren or covered with mere shrubs.

Potential productivity of India's forests has been estimated to vary between 1.8 cu m and 3.4 cu m per hectare per year [Raina, 1982a, p. 7]. Harnessing the potential would require better management of forests aimed at optimum (not necessarily maximum) production. This was necessary not only for meeting the raw material requirements of the forest based industries, but also to enhance fuelwood supplies which formed one of the basic necessities for nearly 80 per cent of the country's population and also for maintaining (if not enhancing) the quality of the environment.

The time is past when forests were plentiful, human and cattle populations were small, and man's needs for timber, firewood, and other forest produce were modest. The nation has already gone through a period of industrial revolution, almost an unprecedented growth in human population resulting in the increasing demand for timber not only for what may be called normal uses such as residential, commercial, and industrial purposes but also for merchant and naval ships, railways, iron ore smelting, etc. These have resulted in a fear of timber famine which, in turn, resulted in the realisation of the need for protecting the forests through prohibitory regu-

[33]Based on a 40 : 60 ratio, this comprises 0.2 sq m timber and 0.3 cu m mixed woods, respectively.

lations. The regulations took different shapes in different countries. In India, the two most important planks were (i) taking over of forest areas under government control, and (ii) regulating their fellings. This marked the beginning of forest management over 100 years ago which primarily consisted of (i) statutes for protecting endangered species, and (ii) practice of the concepts of normal forest and sustained yield.

The term sustained yield was meant to convey undiminishing yield, i.e., managing forests to maintain the existing wealth and out-turn level. Over the years, however, the principle has become a fetish with the forestry system and has prevented the professionals from adopting a more flexible and dynamic approach towards production management. Most countries have already accepted the principle of 'progressively increasing yields' as a replacement for the concept of sustained yield. The time is more than opportune to change from the practice of low investment—conventional/inadequate management—low yield forestry to improved management—high yield—high output systems.

Some beginning in this direction was made through the FDCs. The experiences so far have been mixed.[34] The efforts at enhanced productivity of forest land may have to move at a modest pace for want of adequate financial resources and, more important, for want of technically trained manpower with a weakness and environment for development oriented administration.[35] It may, therefore, be safely said that at least two-thirds of the estimated potential average output of 1.8 cu m per hectare can be harnessed with modest efforts at improved management on at least 55 million hectares of India's forests by the turn of the century. This would mean a total output of 66 million cu m of timber and fuelwood. Afforestation and reforestation of unproductive forest land may necessitate the choice in favour of some fast growing tree species. As a result, the mix of output between timber and fuelwood may change

[34]A detailed assessment of the philosophy and working of the FDCs in India is overdue. It was learnt that a committee was being appointed for this purpose.

[35]Some people appeared to think that the growth of environment for development orientated administration had been stalled by the unprecedented emphasis on conservation in the recent past.

from 40 : 60 to 35 : 65 or even to 33 : 67. In spite of this, timber output would substantially increase. More importantly, however, the total output of materials usable as fuelwood and pulpwood can be raised from the current level at around 30 million cu m to 44 million cu m per annum. On the other hand, the demand by the paper and paperboard industry has been estimated at 2.7 million tonnes or 2.16 million cu m or a maximum of five per cent of the lowest feasible potential.

An important outcome should be the elimination of the current tensions amongst the forestry system, the administrators, the policy makers, and the pulp and paper industry. More importantly, the conflict on the issue of choice between fuelwood and pulpwood should be eliminated. This would be more appropriate with the attempt at afforestation of revenue wastelands, low productivity community lands, rail-road sides, canal sides, etc. through the on-going programmes and projects including social forestry.

3.4.1 Prospects for Enhanced Bamboo Productivity and Production

A number of bamboo species are suitable for pulping. Though its proportionate contribution in the mix of cellulosic raw materials for the industry had declined, yet it was the single most important goods and met nearly 60 per cent of the requirements of forest raw materials. The total consumption of bamboo in 1985 was around 1.7 million tonnes.[36]

Considering the continuing trend towards reduction in the proportion of bamboo in the furnish and the estimated demand of 1.7 million tonnes by the turn of the century, a discussion on the prospects for enhanced bamboo output may not appear to be relevant to this study. On the contrary, one of our main objectives is to discuss the principles which may be used for fixing the administered prices of this produce supplied by the FDs to the industry, and that necessitates an understanding of the industry's demand in relation to total supply.

[36]It may be financially desirable and technically feasible to further reduce this demand for bamboo through development of long fibre and fast growing tree species in some of the high rainfall forest areas.

There is considerable difference regarding the data on potential and actual production of bamboo in the country. Most sources place the annual output at three million tonnes [Bachkheti, 1982, p. 28; Biswas, 1982, p. 136; Ghosh, 1980, p. 7]. On the other hand, at least one estimate places it at 9.5 million tonnes [Vermah and Pant, 1981, p. 466]. Similarly, estimates of natural bamboo growing areas in the country vary from nearly 6.5 million hectares or less than 9 per cent of the total forest land to 10 million hectares or over 13 per cent of the forest area [Biswas, 1982, p. 134; Vermah and Pant, 1981, p. 466]. The estimates of bamboo yield vary between less than 0.5 tonne to nearly 1.0 tonne per hectare a year. These variations show that management of a scarce resource has not received the desired attention.

Let us work with the most common and conservative estimates of output at three million tonnes a year and the natural bamboo growing areas at 6.5 million hectares. Currently, therefore, nearly 1.3 million tonnes should be available for all users other than the paper and paperboard industry.

The tension-ridden situation may, thus, be eased if the total annual output can be gradually raised to reach 4.5—5 0 million tonnes by the turn of the century. This could be feasible from managerial, administrative, financial, and economic angles. The required action may be grouped into three interdependent categories :

—Better management of natural bamboo growing areas;
—Raising man-made plantations of bamboo; and
—Inducing individual land owners to cultivate bamboo not only to meet their requirements but also as a cash crop.

3.4.1.1 *Improved Management of Natural Bamboo Forests*

A conservative estimate of potential annual output of bamboo in India has been around five million tonnes against the actual output of three million tonnes. One can, therefore, say that the current shortage is not due to non-existence of the material but mainly due to inadequate management. Improvement in management practices alone may, therefore, suffice to meet the objective of producing five million tonnes by the turn of the century. Harnessing the whole of the current potential may not be feasible but this part of the action could receive maximum attention.

One relevant fact in this context is that though most of the forests are overexploited, yet there were vast unexploited areas of bamboo resources, mainly due to lack of infrastructural development. These areas were concentrated in the north-eastern parts of the country and the Bastar region in Madhya Pradesh. A 1976 survey of the bamboo potential in Madhya Pradesh, for instance, revealed that annual potential output was over 1.0 million tonnes whereas the actual output was less than 0.4 million tonnes [Singhal, 1982, p. 114]. This situation has not changed significantly. While precise estimates for the north-eastern states were not available, it was generally agreed that the unharnessed portion of the resource in that region could be over one million tonnes a year.[37]

It may, therefore be convenient to think that a single step towards providing approach roads, bridges, and rail links where possible in the exploitable forest areas would lead to enhanced availability of bamboo by a minimum of 0.75 million tonnes a year.[38] At least two questions arise at this stage.

First, should a natural resource be allowed to remain unharnessed for want of infrastructural facilities ? The answer would depend on the relative costs and returns. However, an assessment of the expected returns must also account for (i) retardation in future growth due to congestion and decay due to inadequate harvesting or non-harvesting, and (ii) further enhancement in the pressures on comparatively accessible areas. Such assessments are not easy but our discussions with the concerned people indicated that, on the whole, the investments on opening up a good portion of the inaccessible areas would be recovered within a reasonably short time.

Second, should the investment be made and managed by the forestry system or the user industry ? Arguments for or against entrusting the task to either of the two parties can be made. At the same time, the FDs/state governments who own and manage

[37]Due to abundance of bamboo in these areas, harnessing the resource should not be in conflict with the interests and rights of the local people.

[38]Exact estimates of the harnessable potential from financial, managerial, and environmental angles were not available. The estimate at 0.75 million tonnes, however, was considered to be on the lower side.

the forest land resource have, at least in the past, looked at it more as a source of net revenues. As a result, the gap between revenues from and investments in the forestry system has been rising continuously and the situation may not change in the desired direction at the desired speed.[39]

On the other hand, the industry as the user of the scarce materials may be expected to take up the responsibility or at least share it with the FDS/state governments, and this has long since been an accepted principle. Perusal of the available agreements between some state governments and the industrial units showed that the latter were granted right to harvest bamboo from relatively inaccessible areas at significantly lower royalty rates compared with the royalty rates of the produce from accessible areas. The logic, though not spelled out, appeared to be that the concerned paper mills would invest in the infrastructural development of the inaccessible areas.

Data on the extent of such investments made by the paper mills were not available. Discussions with a number of informed people, however, indicated that investments have not been enough. It has even been observed that yields from most of the bamboo growing areas tied to old leases with paper mills have been substantially lower than the potentials, and the resources in the interior areas have remained grossly underutilised because the lessees have contributed only marginally to their infrastructural development [Sowani, 1982a, p. 39].

The same conclusion may be drawn from another set of happenings. In some cases the state governments have decided to do away with the distinction between accessible and inaccessible areas after over 30 years of supply of bamboo to the existing pulp and paper mills, but the managements of the latter have argued for continuation of the distinction on the grounds that (i) the harvesting and extraction costs from the inaccessible areas are substantially higher, and/or (ii) the invest-

[39]The need for enhanced investments appears to have been recognised at all levels. Some definite steps have also been taken or were proposed to be taken. For instance, Andhra Pradesh Government had proposed setting up a forest development fund. In Maharashtra, a forest development tax has been operative for the last four to five years. But the net impact of such measures on improved management of the forest resource systems remained to be seen or studied.

ments made in opening such areas have not been recovered. Neither side has supported the arguments with facts. The issue could be objectively assessed and resolved if the following data were available :

—Investments made in opening the inaccessible areas,
—Outputs obtained from the inaccessible areas opened in the past and their proportion in the total raw materials,
—Extent of the investment required to harness the remaining inaccessible areas, and
—Annual additional outputs which may be obtained from opening such areas.

The paper mills can provide these data for the past and the Forest Resources Survey Organization (FRS) can generate the data for the future. The marginal cost for the FRS was not expected to be high as the organisation has been preparing fairly detailed inventories of the hitherto unexploited but exploitable forest resources for specified industrial catchments.

Besides inadequate infrastructure, bamboo yields have been lower due to want of appropriate management. The attitudes in this respect may be gauged from the fact that it continued to be classified as a *minor* forest product. It is also known that, for a long time, bamboo management in India was mainly based on trial and error [Qureshi and Deshmukh, 1963, p. 13]. More importantly, bamboo crop has been subjected to overlapping working circles.[40] The management of bamboo has mostly been subordinated to the management of other species. This has even resulted in decreased occurrence of bamboo over large areas.

It could therefore, not be surprising if average annual output of bamboo forest areas in India was less than one tonne per hectare whereas well managed natural forests can yield up to 10 tonnes per hectare a year. This called for changes in the prevalent philosophy and practices. The changes, however, need not necessarily require high investments per unit of land.

[40]There is a difference between the silvicultural requirements of timber species and bamboo. Bamboo not only occurs everywhere, but also occurs as understorey to most of the tree species. Tree crops and bamboo must, therefore, receive different and separate treatments. This could be also the best use of land to obtain two major crops from the same area.

Considerable improvements in productivity at a low cost, particularly from the areas being harnessed, could be possible with marginal investments. We may discuss at least three interconnected steps :

i) Optimum silvicultural practices;

ii) Improved management practices such as gap filling, moisture conservation, fertiliser application ; and

iii) Genetic engineering.

Optimum silvicultural practices were required due to inadequate or inappropriate management of the resources in the past. All concerned appeared to agree that silvicultural practices including optimum harvesting cycles, intensity of harvesting, and methods of harvesting different bamboo species in relation to the rest of the forest crops can be improved.

The second category of improvement measures could consist of gap filling, soil working, water conservation through trenching and mounding, and use of chemical fertilisers. Experiments on these lines in bamboo areas leased to a paper mill in Andhra Pradesh exhibited increase in new bamboo shoots by an average of 54 per cent, and considerable general improvement in the crop condition [Chugh, 1982, p. 65].

Conservatively assuming that : (i) yields as a result of such operations may increase by 30 per cent from the denser bamboo forest areas, and (ii) five million hectares of the denser bamboo areas with current average yield at 1.5 tonnes per hectare can be treated over a period of four years; it has been estimated that outturn of bamboo can be raised by 2.25 million tonnes (0.45 tonnes $\times$ 5 million hectares) within a period of eight years [Chugh, 1982, p. 67]. The industry could undertake the task but would logically expect that administrative arrangements, particularly with respect to royalty rates on the additional outputs, must be finalised in advance.

The concerned government in 1983 made an offer to the paper making units to enhance the supply of bamboo for their own use through improved management of natural bamboo forests already earmarked to them. The main points of the proposal were :

i) The forest land would *not* be given on lease to individual industrial companies but they could improve the management of the area already earmarked to them by gap filling, water conservation, bunding, mounding, chemical fertiliser application, and fencing.

ii) The initial duration of the scheme would be four years but it could be extended by another similar period (*not periods*) with necessary modifications.

iii) The industrial units shall pay for the increased bamboo outputs at the *prevalent royalty rate*, as *revised from time to time*, but a development rebate @ one-third to one-fourth of the development costs subject to a maximum of Rs. 20 per tonne of the bamboo royalty paid during the previous year would be allowed.

iv) The industrial unit (s) would set apart for the FD 20 per cent of the additional produce in any year and the latter would make it available to the general public or to the local people.

v) An industrial unit could seek institutional finance for the proposed improvements in the crop condition.

vi) Only those major companies which agreed to pay the revised royalty rates on bamboo and mixed woods with effect from the earlier *proposed date in 1980-81 could avail of this scheme*.

A number of issues arise from these conditions. At least three issues appear to be more important. First, could the institutions such as NABARD consider financing any plans for better management of forest areas by individual industrial units without their holding the leases for such areas? The answer would be in the negative unless special arrangements between the parties were to be facilitated.

Second, could the industrial units be expected to undertake an activity if the prices of the output were to be revised by the land owner from time to time ? The answer may, once again, be in the negative particularly when there was lack of mutual faith and understanding between the parties, and the general thinking was that the royalties could be revised only upwards.[41]

Third, could the industrial units be expected to agree to pay the royalty rates proposed by the state government starting from 1980-81 when the matter was still pending in the State

[41]A paper mill in Haryana which has been purchasing eucalyptus wood from farmers was reported to have paid 25-30 per cent lower prices during 1986-87 than the preceding years.

High Court and negotiations for an out of court settlement were going on ? The answer was in the negative, and the scheme remained only on paper.

Another example, but on the positive side, was offered by the West Coast Paper Mills who had reportedly done considerable work in the improvement of density of bamboo clumps per hectare as well as in enhancing the production per clump [Sowani, 1982b, p. 57]. The experience needed to be analysed to assess its replicability with or without modifications including administrative arrangements between the state governments/FDs and the paper mills. If mutually acceptable arrangements cannot be worked out, the FDs must undertake the task without further loss of time.

The third measure to enhance the outputs from natural bamboo forests lies through genetic engineering so as to accelerate the growth rate and reduce the gestation cycle. Bamboo was crossed with sugarcane many years ago with the objective of producing a quick growing (annual) bamboo crop. This work has not been carried further either by the FRI, or the Indian Agricultural Research Institute (IARI), or the Indian Council of Agricultural Research (ICAR), or the paper industry [Swaminathan, 1982b, p. 17]. In general, very little research towards genetic improvements in bamboo has been carried out.

The main reason for this situation appeared to be the low propensity to wait for the results which, in turn, could be attributed to the environment which prevailed in the industrial, administrative, policy making, and the professional circles. The environment must change. For the present, however, we cannot estimate the extent of additional bamboo outputs attributable to possible genetic improvements.

The discussion in this sub-section may be summarised to say that bamboo output in India can be enhanced by at least three million tonnes a year within the next 8-10 years by concentrated action towards providing the necessary infrastructure and improving the silvicultural and biological management of the natural bamboo areas.

3.4.1.2 Prospects of Bamboo Plantations

In addition to improved management of natural bamboo areas, financial, technical, and managerial feasibilities of raising bamboo plantations must be adequately assessed. One estimate was that plantations of certain species spaced at 5× 5 metres can yield 15 tonnes per hectare per annum, and the yield can be further enhanced with reduced spacing [Vermah and Pant, 1981, p. 467]. Another estimate has been that bamboo plantations in high rainfall areas with rich soil can yield 20 tonnes per cycle of three years. This meant an average of six to seven tonnes per hectare per year [DCPPI, 1983d, p. 16]. Yet another observation has been that bamboo lends itself excellently for cultivation and this was supported with very attractive figures of IRR, NPV, and return-cost ratios [Sheth, 1982, pp. 58-59].

We do not have the expertise to sort out the significant difference in estimated yields from bamboo plantations. At the same time, financial viability of the activity would largely depend on annual cost and output levels. The degree of certainty with which estimated outputs can be available would be equally important for decision making.

It must also be noted that assessments of financial viability of bamboo plantations must account for savings in harvesting and transportation costs compared with those from mixed forest areas. Such costs for bamboo availed from natural stands have been reported at Rs. 300 a tonne at least in some cases. Those from planted areas are expected to vary between Rs. 150 and 200 a tonne depending on location of the planted areas vis-a-vis the location of the processing units. Plantations would also offer the advantage of uniformity in size and quality of the material and, thus, some savings in processing costs.

There were, however, differences of opinion regarding managerial feasibility of large scale bamboo plantations. On the negative side, biotic interference has been the main issue. Again, it was difficult to adequately analyse this problem. It can, however, be observed that biotic interference is a reality, and the problem may be sorted out by estimating the protection costs as an integral component of the activity. It may also be noted here that the choice of protection techniques would have to be an important consideration. For instance, trench-cum-mound fencing with two to three lines of non-browsable tree species placed on the mounds could offer a low cost alternative in most situations. On the other hand, barbed wire fences can be provided within a short time but the costs per unit of land would be high.[42]

This discussion may be concluded with the observation that bamboo plantations could meet the tests of technical feasibility in most cases, and of financial and managerial feasibilities in selected situations. Moreover, investment requirements per unit of land or per unit of output could be substantially higher compared to the other measures for enhancing productivity of natural forests. Since, capital and technically trained manpower are scarce, it may be desirable to concentrate on improved management of natural bamboo bearing areas. That would also mean a better use of the land resource since bamboo grows as an understorey crop to most timber species.

3.4.1.3 Prospects of Bamboo as a Cash Crop on Private Land

Tree crops have already found an important place amongst on-farm activities in many parts of the country. This has induced a thought for large-scale efforts to introduce bamboo as a cash crop [Biswas, 1982a, p. 136]. It has been observed that rural people grow bamboo in a much better manner than the foresters primarily because they can protect the crop better [Tiwari, 1982b, p. 60]. Farmers in some parts of the country are reported to have taken up bamboo cultivation, but adequate details were not available to enable an analysis of financial, economic and managerial feasibilities of such operations.

It should be feasible to have bamboo plantations along the corners and edges of cultivated fields at least in the high rainfall areas. The NCA estimated that 2.7 million hectares of land available along farm boundaries could be placed under trees. Even if 20 per cent or 0.5 million hectares of this land is placed under well managed bamboo plantations, the pressure on forest land would be considerably reduced.

This implies that policy measures to induce bamboo plantations along the corners and bunds of cultivated fields should be taken up. In the long run, however, the extent of bamboo plantations on individually owned land would be primarily a function of relative profitability of the activity vis-a-vis other crops including tree cultivation. This, in turn, would be a func-

[42]A detailed discussion on the relative merits and demerits of different protection techniques is not within the scope of this study. For details see: Gupta, 1987 b.

tion of : (i) the relative scarcities of the material not only for pulping but also for other commercial and industrial uses, and (ii) the price which the users would be able and willing to pay. Since the behaviour of these variables is difficult to foresee, bamboo plantations on privately owned land have not been visualised to play a major role.

The discussion in this and the preceding two sub-sections would suffice to conclude that: (i) opening up inaccessible forest areas, (ii) improved silvicultural management of natural bamboo forests, and (iii) genetic engineering to introduce fast growing and high yielding bamboo species would suffice not only to meet the demand for the product in the foreseeable future but also to dispense with the cult of tension management.

3.4.2 Prospects of Enhanced Production of Hardwoods

As noted earlier, mixed woods would play a relatively more important role in the category of cellulosic materials for the pulp and paper industry in India in the future. Prior to a discussion on the prospects of enhanced supplies of this material, a few general observations may be made.

1) Contrary to the common belief, which seemed to be primarily due to competing demands on hardwoods for household energy, it has been reasoned that enhanced industrial use of this material may be socially desirable. It has, for instance, been observed that increased use of wood for pulping in the recent past has enhanced its value whereas it would not have found a market even as fuelwood. This, in turn, has encouraged plantations of short rotation tree species and has opened up the possibility of enhancing the financial viability of forestry operations in the country [GOI, 1972, p. 16].[43]

2) The hardwood resources were not evenly distributed in the country. A major portion of the total output was produced in Andhra Pradesh, Uttar Pradesh, Madhya Pradesh, Maharashtra and Karnataka. The short rotation tree species introduced in the wake of increased demand have enhanced the financial fea-

[43]Cases have been noted where at least some FDs and FDCs were faced with the problem of selling mixed woods at reasonable prices during the 1970s, and agreed to supply the materials to paper mills without even formalising the agreements to that effect.

sibility of utilising comparatively less productive forest land and other waste lands in other parts of the country.[44]

3) The trend towards enhanced use of hardwood may be further facilitated by setting up of mother pulp mills wherever the materials are in surplus. This can result not only in better utilisation of the resource but can also reduce the tangible and intangible costs of paper production.

4) The paper mills could be encouraged to introduce marginal changes in the pulping technology to be able to use smaller pieces of pulpwood of which large quantities were left behind in the forests because the mills considered them unpulpable. The main problem, however, appeared to be high cost of transporting these materials. This could be solved with the use of mini chippers or crushers so that the materials were compressed and bundled on the site for ease and economy of transportation [Sowani, 1982a, p. 38].[45]

On the whole, therefore, enhanced use of hardwood including the materials currently considered unpulpable can be expected to go a long way towards scientific management of the land resource, resolving the conflicts amongst competing uses, and relieving the tensions between the forestry system and the paper and paperboards industry.

One stream of thought, however, has been that the future requirements of cellulosic raw materials for paper making would have to be largely met from man-made plantations. This would also facilitate adoption of recent developments in genetic engineering which offer the possibility of phenomenal increase in the biomass output per unit of land [Raina, 1983, p. 5, emphasis added]. This line of thinking has led the pulp and paper industry to demand that it should be allowed to produce the cellulosic raw materials for its own use. Though an assessment of the merits and demerits of such a step is not within the scope of this study, yet it may be noted that the step would be highly capital intensive.

Let us, therefore, proceed with the argument that the pro-

[44]Once again, some people feel that the Forest Conservation Act has given a setback to plantations of fast growing hardwood tree species.

[45]Differences of thought regarding feasibility of mini chippers have already been noted. We do not have the competence to comment on this aspect.

blem of shortage of cellulosic raw materials of forest origin can be solved by simultaneous action on different fronts, and the actions need not necessarily be capital intensive in nature. A few specific steps can be discussed.

First, productivity of forest land can be substantially improved merely by reducing (not eliminating) waste in the harvesting/extraction process. Such waste has been estimated at over 40 per cent in the hills and 20 per cent in the plains [Thapar, 1982b, p. 25]. The objective could be achieved with marginal improvements in harvesting technology and techniques.

Similarly, destumping in the mixed forests clearfelled for man-made plantations should be a cost effective measure to raise the outputs of mixed woods per unit of such areas. One must, of course, think of environmental impacts of such measures. Discussions with knowledgeable people, however, indicated that environmental considerations of monoculture would usually not vary 'with' and 'without' destumping the natural crop. It has also been argued that the practice of clearfelling mixed forest to raise commercially valuable crops of single tree species (even on a limited scale) should be either discarded or all out efforts should be made to induce the growth of man-made crops. In the latter case, destumping could even be desirable as the soil would be worked up in the process.

Second, a simple augur method of tree planting in saline and alkaline soils is reported to have been developed. By using this technology, gypsum does not have to be spread all over the plantation area. Only small quantities of gypsum need be placed in pits to augment survival and growth of trees [Swaminathan, 1982, p. 16]. This, if feasible, can substantially enhance the possibilities for covering the vast extent of wasteland in and outside the forest areas.

Third, species such as *Moringa oleifera* (Sanjana/shevgi/sargava/bhendi) which (i) can be raised by planting branches, (ii) can easily attain a height of five metres in one year, and (iii) can be pollarded every year starting from second year of planting, could be popularised to enhance pulpwood production. The species can adapt to most situations except in the Himalayan region, the north-eastern region, and the Thar desert [Tiwari, 1982a, pp. 52-54]. *Sesbania grandiflora* (Agusth) is another tree species in this category. It is considered to be particularly suitable for southern and central regions of India. Its flowers are edible, leaves are good as fodder, and can be worked on a five year rotation [*Ibid*, p. 54]. It can also grow on partially saline soils which may be better developed to enhance pulpwood outputs than to raise seasonal crops. In short, many tree species offer low cost alternatives to meet the objective.

Fourth, eucalyptus has occupied a prominent place as a source of raw material for pulp and paper industries in many countries. It is a fast growing species and a good yielder. Most eucalyptus species possess properties such as light coloured wood, low basic density, and low percentage of extractines. These qualities are desirable for pulping. During the last two decades large scale plantation of eucalyptus have been taken up in many parts of India. With a 8 to 10 year rotation, the average yield per hectare per annum could be as high as 30 tonnes [Anon, 1965, p. 63]. Some species can yield over 40 tonnes per hectare per annum.

Capital requirements of eucalyptus plantations have been substantially lower compared to the broad leaved species. More importantly, the crop need not necessarily be raised as plantation of the single species. Average yields at 10 tonnes per hectare per annum were not unknown even from mixed forests. It should also not be necessary to pulp the entire tree. A good portion of the main trunk can be used as timber (large and small) while the upper one-third or so can be pulped. It may also be noted that, barring highly commercially oriented situations, eucalyptus need not occupy prime forest or farm land. In the Kutch area, for instance, the farmers had taken up tree plantations in a big way and eucalyptus was one of the popular though not prominent species.

Fifth, irrigation water need not be extensively used for raising tree crops. We do not visualise much use of either the surface and ground water systems created for seasonal crops. However, some other measures such as soil conservation in hilly areas can substantially enhance the moisture regime and, thus, enhance forest outputs.

Moreover, effluent water discharged by industrial units is considered usable for irrigating tree crops. This may be particularly feasible in the vicinity of paper mills, may reduce the rotation age of tree species to one-half or even one-third of the normal time, and also result in increased outputs per rotation. The task may be facilitated through supply of effluent water for limited, say two to three, irrigations per year. The pulp and paper producers may also not continue to insist on eucalyptus, and may even encourage the farmers to raise multi-purpose tree species.

Sixth, supplies of cellulosic raw materials of forest origin can also be enhanced through vertical integration of some of the wood production and processing functions. A portion of timber logs sold at forest depots could, for instance, be converted into finished items such as doors and windows, furniture and fixtures. The FDs have a distinct advantage in undertaking such activities. The consumers would also benefit especially in terms of assurances regarding the quality and the species of wood used. The activity would, of course, have to be undertaken on a reasonably large scale to be able to minimise overhead costs. Cuttings and sawings accumulated at centralised places can be made available both for fuelwood and pulping [Maslekar, 1980, p. 191].

Seventh, and more important, the term 'plantation' must not be interpreted in its literal and limited sense to mean placing nursery raised seedlings in pits and/or mounds. One consequence of such limited interpretation has been that administrative, financial, economic, and managerial costs per unit of additional outputs from enhancing the growth of naturally occurring tree species have not been assessed [Gupta, 1987c]. Similarly, the practice of raising trees by sowing has also not received the desired attention. These can be important steps to speedily enhance the productivity of the wastelands at relatively low costs.

It must also be noted in this context that vast tracts of forest and other common land in the country remain devoid of adequate cover almost round the year. Some grasses and shrubs come up during the rainy season but do not last long as they are consumed by herds of cattle, goats, sheep, etc. The land thus overused and misused has been continually deteriorating both physically and chemically. Heavy soil working to house nursery raised plants in such sites may neither be financially feasible nor technically desirable. An immediate requirement is to adequately cover the land to enable the soil to regain a good portion of the lost nutrients and shed some of the undesirable features such as high pH. This may be done within a short span of five to seven years, and the wastelands can turn to be productive lands.

These considerations coupled with the need to contain the investment requirements could lead to the prospects of harnessing the established root system of species such as *Acacia arabica* (Desi Babool) and *Prosopis juliflora* (Vilayati Babool). Wild growth of Vilayati Babool can be periodically thinned to encourage the growth of one or two relatively healthy and straight stems per plant which can grow into trees within a period of four to five years. The weaned materials can be used as fuel. The essential point is that tree species which occur naturally on a considerable scale in most degraded areas, possess good coppicing power, and can also be raised by direct sowing; hold the potential for highest biomass production per unit of land and time, and the lowest possible capital and O and M costs [Gupta, 1987a and 1987c].

It should also be interesting to observe at this stage that (i) *Acacia arabica* was one of the hardwood species used by the Sheshasayee Paper Mills along with eucalyptus, and (ii) young (three to four years of age) *Prosopis juliflora* was placed in the first grade of hardwoods from the point of view of its suitability for pulping [Misra, 1973 and 1978].

3.5 Some Policy Implications of Harnessing the Prospects of Enhanced Outputs of Bamboo and Mixed Woods

The discussion in this chapter should suffice to conclude that a mixture of conventional and unconventional measures for planning and management of the forest and non-forest land resource can be used to meet the objective of enhanced outputs of timber, mixed woods, and bamboo within the next eight to ten years; and most of the chosen measures need not be capital intensive.[46] Between bamboo and hardwoods, the prospects of raising the outputs of the latter may be substantially better but

[46]Financial resources for forest development may not be as severe a constraint as was the case till the recent past.

the industry can be expected to gradually change the raw materials mix.

The task of supplying adequate quantities of bamboo and mixed woods to the industry can be further facilitated if other forest-based material such as grasses and reeds could support about 10 per cent of the total outputs required by the paper industry in India by the turn of the century. We have, however, been conservative in this regard as the administrative and marginal feasibilities of large scale use of such materials remain to be studied; and there could be a relative advantage in using them for newsprint production (not covered in this study) than for the production of paper and paperboards.

There is no need to emphasise here that the long-run practices and policy for pricing (not royalty) the industrial raw materials would and should be a function of the degree of success of the measures for their enhanced outputs, and the task of assessing the policy options can be objectively approached with the success of the desired and feasible efforts. On the other hand, however, some unconventional alternatives may have to be chosen if all concerned are not willing to face the challenge and cease the opportunities.

One of the thoughts has been to raise plantations to meet the raw material requirements of specific industries. Based on studies conducted by the NCA and the FRS, one estimate has been that a 1,500 square km of adequately stocked forest area can perpetually feed a 300 tonnes per day (say 1,00,000 tonnes a year) capacity paper mill [Raina, 1982a, p. 17]. This is a liberal estimate as it implies an average annual output of only 1.7 to 1.75 tonnes per hectare. Even this, however, means that 1.5 million hectares or 2.5 per cent of the currently managed and potentially manageable forest land can meet the pulp and paper industry's demand for mixed woods.[47]

Assuming a yield from industrial plantations at 10 cu m or eight tonnes per hectare per year, another estimate of land required to feed paper mills with capacity at 400 tonnes per day or 1,30,000 tonnes per annum has been 70,000 hectares [Agarwala, 1982, p. 45]. The arithmetic is not clear. Moreover,

[47]Bamboo supplies to the extent of 2.5 million tonnes could be conveniently available to the industry from natural forests.

according to some senior forest officers, an average output of eight tonnes per hectare a year may not be feasible as the land available may be poor and allocation of forest areas exclusively for pulpwood may be difficult. But, even if the output from the plantations could be six tonnes per hectare per year, the total area required to meet the paper and paperboard industry's demand for woody materials by the turn of the century should be less than 0.75 million hectares or around 1.3 per cent of the productive forest area in the country. If it could be assumed that bamboo would continue to be available from natural forests, the land requirements for hardwood would be only 0.44 million hectares or about 0.8 per cent of the productive forest area.

These differences in requirements of land for industrial plantations should serve to emphasise that any one estimate of average output per hectare per year would not suffice. The yield levels would vary according to agro-climatic conditions and the chosen tree species and may, therefore, have to be worked out for individual paper mills. There are at least two other important and connected issues. First, man-made plantations need not be raised to feed all the existing and upcoming paper and paperboard making units. Second, all of the outputs from the industrial plantations need not, or even should not, be used for pulping. The manner in which timber and poles are marketed and/or processed would have definite bearing on land requirements for raising industrial plantations, and on pricing the materials used for pulping.

An equally important issue would be whether industrial plantations should be raised only on forest land. It has, for instance, been estimated that two rows of trees on both sides of a road or rail track of one running km would equal one hectare of plantations. The length of the country's road and rail tracks is over 2.0 million km. If 25 to 30 per cent of the road and rail road sides can be appropriately planted for the purpose, the industry's demand for wood by the year A.D. 2000 may be fully met. The main point is that the demand for industrial raw materials can enhance the feasibility of placing the waste lands under tree crops.

The pulp and paper industry has also been keen to raise 'captive' plantations. The argument has been that such plantations would provide a sustained and cost effective solution to the problem of scarcity of cellulosic raw materials. The materials thus produced could also be stored and processed at relatively lower costs.

These arguments have not found much favour with the policy makers. The prevailing socio-political dynamics have been a major consideration. If, however, the forestry system does not consider it feasible to raise the productivity of the land resource under its control and management, and if a time bound programme to cover the wastelands cannot be drawn and implemented by the agencies considered appropriate for that purpose; administrative, managerial, and financial feasibilities of leasing a portion of the forest and/or other common property land to the industrial units to raise bamboo and tree crops must be examined objectively.

CHAPTER 4

Bases for Pricing Cellulosic Raw Materials of Forest Origin : An Assessment

Pricing forest produce was not an important issue till recently as the supply was considered to be almost inexhaustible. Apart from the age old practices of open auctions or inviting sealed tenders for a large portion of the output including highly valuable timbers, and high level negotiations for fixing the royalty of materials supplied to the forest-based industries, there have been very little efforts at systematic marketing wood and non-wood products.

Almost all the FDs have had a Forest Utilisation Officer (FUO) of the rank of Divisional Forest Officer (DFO). He was expected to look after the marketing and pricing activities, and to maintain liaison with the forest-based industries and other users of forest products so as to *create demand* for them. The FUO's office was, therefore, considered to be the centre for all the market related information : quantitative and qualitative. The FUO's counterpart with the FDCs was usually designated as Regional Manager Marketing (RMM).

These offices have been ineffective in supplying data on the quantity of different forest products sold or the prices obtained at various sale depots; quantity of materials sold to different categories of users, such as forest-based industries, saw millers, furniture manufacturers, charcoal makers; ruling market prices for different types and grades of timbers, and other forest products in the country; and consumers' preferences for varying timber sizes. The data were, of course, collected but redundant by the time they were sorted and entered in the prescribed registers [Maslekar, 1980, p. 195]. In spite of this, the concern for pricing the administered supplies of the produce to forest-based industries has been apparent.

It has been already discussed (Chapter 2) that the paper industry in India has had a long and chequered history from a number of angles including the mix of cellulosic materials used for pulping. The most common of these till the 1920s were sabai grass, hemp, rags, jute, imported spruce, wood pulp, etc. Bamboo came to occupy the most prominent place during the 1930s when it was looked upon as a wild grass and was required to be taken out of the forest areas to enhance other tree growth. Those concerned including policy makers, administrators, and professional foresters felt that whatever could be received from the bamboo users was a *net gain*. Therefore, the concept of cost of production for bamboo was not relevant. This would also explain why royalty was not intended to be a price in the economic sense of the term.

Relationships between the state governments and paper making companies were governed by formal agreements which specified the quantity of the material to be harvested from specified forest areas, for specific period of time, and at specified royalty rates. Initially, the royalty rates were nominal and there was hardly any thought given to finding a logical basis for determining the rates. The situation, however, has changed during the last 10 years or so. The policy makers, administrators, foresters, academicians, and others concerned have emphasised that adhocism in fixing royalty rates must yield place to rates based on more scientific reasoning. The industrialists have also exhibited sufficient interest in the issue.

Some reasons for the need for change in thinking have been discussed in Chapter 2. It must also be added that the issue of pricing forest-based industrial raw materials is relatively more important in India compared to the developed and free market economies where development of wood-based industries have been supported by plantation of tree crops. In India, on the contrary, the FDs control and manage over 95 per cent of the total forest area and continue to be responsible for the production of a significantly large portion of total cellulosic materials. On the other hand, most of the processing activities continued to rest in the private sector. This introduced a dichotomy in the system.

In the context of their history and the organisational structure, the FDs must also meet some other social and economic obligations : requirements (not demand) of wood and non-wood materials of at least the people directly dependent on and living close to the forests, and adequate supplies and reasonable price for fuelwood for household uses. The task is not difficult if the productive forest land in India could be managed so as to produce two-thirds of the lowest of the estimated potential output during the next 15-20 years (Chapter 3).

Recognising the imperativeness of forests as a major source of cellulosic materials for the paper and paperboard industry and the competing demands on them, the Central Government has been considering issuing of licences for new large and medium sized production units only if the concerned state government/FDs assure sustained supply of the required quantity of these materials.[48] Decisions to set up industrial units must, of course, be taken following appropriate assessments of technical, financial, administrative, and managerial feasibilities. The essential and starting points in such exercises have to be assured supplies of all inputs, availability of a viable processing technology, estimated costs of the inputs, adequate demand for the outputs, and estimated price of the outputs.

These enabled the industry to argue that periodic increase in the price of raw materials and uncertainties in their availability can completely upset their assessments of financial viability. That would, however, be more true for inadequate supply of inputs rather than for the rise in their prices. The output prices may also rise over time, and the rise may even be at a faster rate than the rise in input prices. More importantly, appropriate and desired changes in production technology can enhance the industry's capacity to absorb higher raw material prices. A reference to the initial/administered price of raw materials may, therefore, not be relevant particularly after the production activity has paid back the initial investments.

This implied that (i) assured supplies of adequate quantities of primary raw materials for a specified period of time, (ii) the price at which raw materials would or can be supplied initially and its basis, and (iii) the frequency with which raw material

[48]This was not always so.

prices can or may have to be revised and the basis or bases for arriving at revised prices were inter-connected issues.

Therefore, the main objective of this chapter is to assess a number of suggested bases for determination and periodic revision of price of these raw materials and to attempt a synthesis among these bases in the hope of resolving the current conflicts. To start with, a brief description of the thinking on the subject at the GOI level must be useful.

4.1 Thinking at the Government of India Level on Determining Price of Forest Produce Used as Industrial Raw Materials

It has been observed earlier that the question of evolving a scientific system for fixing the price of raw materials supplied to forest-based industries has been engaging the attention of the state governments and the GOI at least since 1976 when the NCA submitted its report (for details see pp. 82-98). A sub-committee appointed by the GOI to study the issue also submitted its report in 1976 or 1977. Following this, the Central Board of Forestry (CBF) invited the state governments suggestions on the subject. The response was, however, poor in spite of a clear recognition of the importance of the issue.

Subsequently, the CBF appointed another sub-committee. The starting points for this sub-committee's deliberations were :

—Adhocism prevailed in fixing the prices of forest-based raw materials used by the paper industry in India.

—Forestry was a *quasi-commercial* sector and was, therefore, entitled to an adequate return on the investment without which its developmental activities have been seriously hampered.

—The forestry sector should legitimately be assured of the market price of the produce and that price should not be less than the cost of production including a reasonable profit.

—The industries are dependent on the forestry system for prime raw materials and should, therefore, pay reasonable price to sustain the interest of the forestry system and thereby enable it to carry out the needed maintenance and improvement in natural forest, and to raise commercially viable man-made plantations.

—The royalty on the raw materials was a negligible portion of the price of the finished product and the industry should, therefore, not grudge any attempt on the part of the forestry sector to get adequate returns on their investments.

—The FDCs had to work with borrowed capital. They should, therefore, not only be able to recover the capital with interest but should also earn an adequate net return on the investment [GOI, 1978 b].

It can be observed that these statements touched on a number of issues and are repetitive. Some of these were even contrary to the facts/realities. For instance, the statement that forestry development has been hampered due to inadequate returns on the investment cannot be valid as the investment per hectare of forest land in India has been extremely low, returns per unit of investment have been high, and the gap between revenues from and planned expenditure on forestry activities has been continually rising over time [Gupta, 1978].[49] On the contrary, low outputs per unit of the forest land could be largely attributed to the management practices in vogue including low investments per unit of the land resource.

The main thrust of these statements, however, appeared to be that price of the produce from production forests should be sufficient to pay for the cost of clear-felling, raising and maintenance of plantations, harvesting, etc., and to generate reasonable profit. This was in line with the NCA's thinking. Moreover, the sub-committee recognised some general principles evolved by an earlier committee on the subject. These were that prices of forest-based industrial raw materials could be :

—Fixed on the basis of stumpage price which, in turn, could be determined by working backwards from the market price paid by other users of such materials.

—Determined by assessing the delivered cost of the raw material at mill sites including costs of extraction, transport, etc.

[49]The situation did not appear to have substantially changed in the recent past.

—Revised occasionally considering the rate of inflation during the intervening period.

It would be important to observe that the term *price* was used in specifying these three principles.

The sub-committee, however, suggested yet another principle whereby royalty of the produce from natural forests could be determined on the basis of costs incurred in administration, protection, improvement, etc. of such forests; and further argued that any one of these principles could not be adopted by excluding others. They, therefore, classified the problem into two distinct categories:

A) Pricing the produce from man-made forests

It was argued that the price of product from man-made forests should be based on the prevailing market price of the produce and that should not be less than the cost of production including land rent, interest on capital at the normal bank rate, and at least 10 per cent net surplus on investment. This would be discussed in detail at a later stage in this chapter. It must, however, be noted here that land rent must be a component of the cost of production but guidelines for determining this component were not suggested. There are number of problems regarding this point.

One can argue that the rent on forest land could be equal to the lease rent charged by the FDs from the FDCs. However, it cannot be representative and would be an insignificant portion of the costs per unit of output. In most cases, the FDCs had been allocated highly productive forest lands whereas the lease rent charged rarely exceeded Rs. 20-25 per hectare per annum. The main reason for this situation was that the FDCs have been government undertakings and the amount of land rent payable by them has been a non-issue. The charge was levied merely to signify that the FDs/the state governments own the land. This basis would be all the more difficult to practise where the FDCs worked as agents of the state governments and matters relating to fixation of lease rent payable by them were pending for long.

In economic parlance, any unit of a factor of production earns rent when the income received by it is higher than the minimum amount of money necessary to induce that factor to stay with its occupation under consideration. In that sense, the surplus receipts over the minimum supply price of a production agent are termed as rent. Such surpluses would be earned whenever and wherever the supply of a factor of production is less than perfectly elastic, i.e., it is scarce. In spite of its scarcity and high opportunity cost, however, this concept cannot be applicable to forest land in India because of the legal constraints on its alternative uses.

One approach to solve this problem could be the licence fees which the non-government organisations (NGOs) and individuals may be willing to pay for using forest land for raising tree crops. But a representative figure can be arrived at only after :

—It is accepted as a policy that forest lands can be licenced out to NGOs, and other private organisations and individuals for purposes of raising tree crops for their own use or for the market ; and

—A sizeable number of such organisations and individuals come forward to seek licences for such uses of forest land.

B) Pricing the produce from natural forests

The sub-committee suggested that the royalty rates for the produce from natural forests used as industrial raw materials should be fixed on the basis of the prevailing market price of the produce subject to a *minimum equal to the replacement costs* which, in turn, must account for land rent, and costs of administration, protection, regeneration, development and maintenance of infrastructure, etc., and cost of other investments at the normal bank rate of interest.

Once again, there would be many problems in putting this thought into practice over and above the problems in determining land rent. For instance, the cost of protection has to be an important function of the prevailing socio-political dynamics which would largely determine the extent to which the offenders are penalised. Similarly, the regeneration costs may be negligible if the forces of nature are given a free play and appropriate site specific silvicultural practices are followed. On the contrary, the regeneration costs could be considerable if management practices such as soil working, water conservation, and fertiliser application are adopted. Even in such cases, the costs per unit of output need not necessarily be high. On the contrary, the costs per unit of output may even decline if the marginal increments in yield are higher than the marginal costs.

A major problem would also arise, in fact has arisen, from the use of the term 'replacement costs' in place of 'regeneration costs'. The two cannot have the same meaning and policy implications. The sub-committee, however, appeared to be referring to the latter. In both situations, the sub-committee recognised the need for revision in royalty rates every three years. The suggested basis for such revisions was the trend of wholesale prices of the forest produce or of other comparable commodities without the constraint of any ceiling on the revised price.

4.1.1 Some Observations on Government's Thinking

Besides the problems discussed earlier regarding practicability of the sub-committee's suggestions, a few additional observations could be made on the issue.

1) Two seemingly different bases for pricing the produce from man-made and natural forests were suggested but market price has been a reference point in both cases.

2) Terms such as royalty and price, production costs and replacement costs were used interchangeably whereas their meanings cannot synonymous in the context of man-made and natural forests.

3) Revision in prices/royalty rates was suggested every three years without any constraint or ceiling on the revised price/royalty rates. This may be logical but not practical. The buyers' perceptions regarding the price of a good and their capacity to pay that price have to be some of the most important considerations for a price fixation exercise in general, and fixing administered prices in particular.

4) Trend in wholesale price of forest produce under consideration or of other comparable commodities were suggested as the basis for periodic revisions in prices/royalty rates. The implications of using any one of these bases would most likely be different from using the other which can result in an endless debate between the buyers and sellers of the produce. The

debate for identifying a 'comparable' commodity for the purpose can also be long and inconclusive.

The discussion in this section can be concluded by saying that the thinking on the subject at the GOI's level or with the CBF has been far from clear, and those responsible for the job have not paid sufficient attention.

4.2 Assessment of Approaches for Pricing Forest-based Cellulosic Materials

A number of approaches for pricing of the cellulosic materials of forest origin and for periodic revisions of such prices have been suggested by committees of the governments, public and private organisations, and individuals. These include:

—Some past and current royalty rate(s)
—Payment of rent of the forest land
—General price index
—Capacity of the paper and paperboards industry to pay
—*Some* share in profits of the paper producers
—*Some* portion of the prices of final products
—Market price of the goods under consideration or of comparable goods
—Cost of production or replacement costs.

We would assess the logic and feasibilities of each of these in the next eight sub-sections which will help in identifying a logical and feasible basis or a set of bases to resolve the conflicts between the forestry system and the government on the one hand and the paper and paperboards industry on the other.

The discussion would be confined to pricing of bamboo and hardwoods originating from forests as these (along with used paper) are likely to play the most important role to feed the paper industry in India during the next 15-20 years.[50] This should not mean that agricultural and agro-industrial residues cannot or would not play an important role as sources of cellulosic raw materials. The prices of these materials used for pulping were determined on a one to one basis by the producers

[50]The indigenous market for waste paper is well organised and appears to call for no further discussion. Similarly, the current policy with respect to import of waste paper is adequately clear.

and/or dealers of these goods, and the pulp and paper making units.

4.2.1 Pricing on the Basis of Some Past or Current Royalty Rates

Some people are of the opinion that the price of cellulosic raw materials can be fixed by a state government at a given time by taking into consideration (i) some past and/or current royalty rates in the same state, and/or in the neighbouring states, and (ii) the general rise in prices during the invervening period.

This could be a relevant basis for negotiations between the state governments and the newly established pulp and paper making companies for revising the existing royalty rates till the distinction between royalty and price is adequately appreciated. A scanning of the current environment indicated that a major source of the problems in the recent past has been the difference in the paper and paperboard manufacturers' perceptions of the extent to which the royalty rates may be revised upwards and the revised rates proposed by the governments. Since the perceptions or expectations are largely based on experiences, a total and sudden breakaway from the past may not be desirable.

At the same time, the suggested approach cannot be used as a basis for pricing the forest produce used by the paper and paperboards industry. It implies that the royalty rate to be chosen to serve as the basis was logically and/or objectively determined, but that has not been the case. Moreover, the royalty rates have varied considerably not only amongst states but also within the states. Some of the reasons for this have been varying distances between the processing units and the forest areas, relative ease (or difficulty) and consequent differences in cost of harvesting the material, etc. but a definite correlation between these variables and differences in royalty rates could not be found.

It could also be difficult if not impossible to find the concerned parties in agreement on the choice of commodities for computing the changes in price during the intervening period. Weights to be assigned to the chosen commodities would be equally difficult.

4.2.2 Price Based on Rental Value of Forest Land

Another thinking has been that the price of forest produce must equal the land rent. This is the minimum that the forestry system must expect for the produce from natural forests. The industry can, of course, not think of anything more favourable, but the principle is neither logical nor practical for a number of reasons.

1) Price of the produce even from natural forests must pay for not only the land rent but also for other costs on account of protection, normal management, administration, etc.

2) Selling the produce from natural forests at a rate equal to the rental value of the forest land would amount to going back to the concept of royalty whereas the objective is to find a basis for pricing the produce from natural as well as man-made forests.

3) Rent of any piece of forest land would be a function of the species raised. In the foresters' dictionary this is known as the soil expectation value and is determined with the use of Fausman's formula. This, however, is not the relevent concept of rent for professional economists and managers. The problems of determining an economic rent for the forest land have already been discussed (section 4.1). It should suffice to repeat that land rent cannot be a relevant consideration till such time when:

—The policy to lease out forest land to the NGOs/private organisations/individuals is accepted, and

—A sizeable number of such organisations and individuals come forward to seek licences to use forest land to raise plantation crops for their own use or for the market.

Indeed, there could be many more variables which may go into the determination of a long-term pricing policy.

4.2.3 Prices Based on General Price Index

It has also been suggested that the price of forest-based materials should be linked with the general price index or with the overall cost of living index. It may be obvious that such indices cannot be the basis for initially fixing the administered price of any good. One can think of their use for periodic revisions in the administered prices, but that may also not be logical and practicable for a number of reasons:

1) There is an inherent assumption that the initial price was objectively determined but that, as discussed earlier, has not been true.

2) The choice between the general price index and the cost of living index can be a source of debate between the concerned parties which, in turn, may interfere with or impede the discussions on more substantive issues. More importantly, such indices are known to fluctuate considerably from time to time/year to year, and prices of forest products do not bear much relationship with the prices of some of the most important consumer goods like edible oils, foodgrains, textiles which command higher weights.

3) Even within forest products, prices of cellulosic raw materials cannot be linked with those of, say, timber. Demand for the latter is derived from a different set of variables which do not necessarily have a strong relationship with the factors affecting the demand for paper and paperboards.

It is also known that at least till the middle of the 1970s timber prices fluctuated considerably and that situation need not be ruled out for the future particularly if the efforts to enhance timber output through plantation forestry succeed to the expected extent, and work for finding and developing acceptable substitutes for timber continues to progress. It can even be argued that the environment for forestry development in India may not be favourable till the prices of forest products are stabilised.

4.2.4 Prices Based on the Producers' Capacity to Pay

There have been suggestions that the royalty rate should be fixed at the inception of a new paper and paperboard making unit and the rate so fixed should be commensurate with what the industrial unit can afford to pay after ensuring a reasonable profit for itself. The reasoning is that the royalty rate should be attractive to induce industrial investment. This reasoning, however, cannot be valid for at least three reasons:

1) In spite of adequate care and close scrutiny, the initial financial projections made by a firm are liable to change in the long run. The possibility cannot be ruled out even for short

term. The reasons for such changes were known to professionals and practising managers. The important changes in the context of this study could be on:

—Excise duties and other taxes on the product(s).
—Export/import policies.
—Availability and price of substitutes.
—Consumers preference for the substitutes.
—Structure of national and international markets.
—Production technology.

2) It has been observed that royalty of cellulosic materials accounted for a very small part of total cost of pulp and paper production. A major portion (up to 90 per cent in some cases) of the total cost of cellulosic materials was incurred on their harvesting, bundling, transportation, stacking, and storage. Marginal improvements in managing these activities can substantially reduce the cost of cellulosic raw material in spite of a higher *in situ* royalty or price.

3) Inputs such as chemicals, coal, electricity, oil, and water also formed important components of the cost of production of final goods. The industry argued that its capacity to pay higher prices for the forest-based raw materials had been reduced by the spiralling prices of other inputs. But, there was considerable scope for reducing the quantities of these inputs per unit of the final outputs.

For example, consumption of caustic soda varied between 39 kg and 141 kg per tonne of white printing paper among various mills in the country. Similarly, the consumption of chlorine varied between 33 kg and 150 kg [Raina, 1982a, p. 9]. It has also been estimated that the consumption of electrical energy per tonne of paper varied between 1,290 and 1,985 KWH amongst the paper mills in India [IPMA, 1982, p. 14].

In the same vein, it could be noted that even if water is not a major cost item as most of it is drawn from rivers for negligible charges levied by the state governments, its consumption per tonne of bleached paper varied from 250 cu m to 400 cu m. These variations have been largely due to differences in efficiency and the extent of re-use of the resources at various stages of the production process.[51]

[51]A pulp and paper mill with a closed backwater system can be operated with 200 to 250 cubic m per tonne of output [Mehta, 1981, p. 17].

These examples show that costs on account of chemicals, coal, power, water pumping, etc. can be substantially reduced. Similar steps should also reduce the problem of effluent discharge and, thus, help in preserving the quality of the environment. Considerable research and development work at industry and institutional levels is required. Financial resources allocated for the purpose can be more than recovered in a short span. But the relative shares of the different entities in raising these resources has been a ticklish issue.

Manufacturing units which have long since reached the stage of zero depreciated value must contribute relatively more to the R and D efforts. Another area for further discussion could be to charge proportionately higher prices from those manufacturing units which consume relatively more chemicals/power/water, etc. Some norms per unit of final output would have to be fixed and that should not be difficult.

4.2.5 Price as 'A' Portion of Cost of Paper Production

A so-called simple formula such as:

DCM = R/P + E + T

Where:

DCM = Cost of the cellulosic materials delivered at mill sites

R/P = Royalty or price *in situ*

E = Costs of harvesting and extraction of the materials including overheads

T = Transport costs including overheads,

has been suggested for the purpose.

This formula considers DCM as a fixed portion of the cost of paper production which must be determined at the initial stage of determining and/or revising the *in situ* price/royalty rates, and after considering representative data for E and T. In other words, the price can be determined once—

—The portion of the cost of cellulosic materials in the cost of final product(s) is agreed, and

—Representative or acceptable estimated data on extraction and transport costs including overheads are available.

This could be considered logical. An advantage of this principle would be that the *in situ* price of the materials from the accessible forest areas, and/or areas closer to the processing units would be higher and vice versa. But, this also means that the *in situ* price should either be reduced with increase in extraction, transport and overhead costs; or the forestry system must convince the paper making companies to bear a higher proportion of the cost of woody materials in the total cost of final outputs.

The problems in practising this principle can be easily visualised and are, therefore, not discussed further. Some of the major concerns could be:

—The basis of arriving at a standard cost of production of final outputs.

—The prospects of fixing an agreed portion of the total cost of final outputs as the cost of cellulosic materials.

—The basis for arriving at representative/standard costs for harvesting a standing crop, and collecting and transporting the materials to the mill yards.

—Capacity of the forestry system to control the harvesting and transportation costs once fixed;

Or

—Capacity of the forestry system to convince the paper manufacturing companies to absorb a higher proportion of costs on account of the cellulosic raw materials.

4.2.6 Price as 'A' Portion of Final Product Prices

Organisations such as the Development Council for Pulp and Paper and Allied Industries (DCPPI) and some professionals have suggested that the prices of woody materials used by the industry can be floated so as to form a percentage of the prices of final outputs. The suggestions for such prices *in situ* have varied from one to three per cent [Kothari, 1971, p. 293; Sagreiya, 1971, p. 558]. This may sound logical but it did not appear to be administratively and managerially feasible for a number of reasons:

1) It would be extremely difficult to decide on a figure or percentage which the producers and processors of woody materials would consider as reasonable or justifiable basis for fixing

the price. Moreover, the age old tradition of time consuming rhetoric could continue with every revision.

2) The pulp and paper manufacturers have to be more concerned with landed costs of the materials than their price *in situ*. At the same time, costs on account of harvesting, handling and transportation did and have to differ due to differences in the location of paper mills vis-a-vis forest areas, differences in infrastructural development in the forests areas, etc. The situation can be used to enhance the prevailing confusion by those who may derive some short-term benefits.

3) There are significant variations amongst paper and paperboard making units with respect to (i) product mix, and (ii) the input mix including non-woody cellulosic materials. The policy objective has been to encourage the use of agricultural and agro-industrial residues, waste/used paper, etc.[52] Quality of the products and their prices would be a function of the type and quality, and the mix of the cellulosic materials used in the process.[53] More importantly, equity in the prices per unit of different types of these materials would be required but has been and would be difficult to attain.

4) Determining quantities and prices of different types of paper and paperboards produced by each paper making unit would not be easy. Alternatively, some kind of norms can be used but those would also be difficult to determine for reasons discussed earlier.

4.2.7 Price as 'A' Share in the Paper Producers' Profits

Price of woody materials used for pulping can be so fixed that the state governments/FDs as producers and suppliers of these materials receive a predetermined share in the profits accrued to the paper and paperboard manufacturers. This thought appeared to have originated from an interpretation of royalty (not price) as the right or prerogative of the govern-

[52]Enhanced use of each of these materials may have different socio-economic, administrative, and managerial implications. An assessment of policy interventions is, however, not within the scope of this study.

[53]Quality of output would also be a function of the type of plant, machinery, equipment, etc. but such variables are not relevant to the point under discussion.

ments who as owners of the land resource can claim a share in the natural produce from the land in return for allowing other organisations or persons to use the asset(s). This, as stated earlier, means that royalty can be fixed only on naturally occurring produce from natural property belonging to a government [Waheedkhan, 1971, p. 559].

It has been further reasoned that the natural produce for which royalty can be fixed may be divided into two broad categories :

—Product from those natural assets (such as forests) which are almost unutilisable or those where the government is not prepared to make adequate investments for development because of limited demand for their produce.

—Naturally occurring produce *needed* (not necessarily demanded) by a majority of the people [Waheedkhan, 1971, p. 559, emphasis added].

In that sense, the royalty rate of these categories of natural produce should have no connection with their market prices but should be fixed on considerations such as living standards of the people, and/or the stage of industrial development of a nation/state, if the produce is so utilised [Waheedkhan, 1971, p. 560].[54] This logic was used to argue that the royalty for pulpwood supplied to the paper and paperboard manufacturing firms in the country can be fixed on varying standards with varying situations. The reasoning was explained with three examples:

1) In forests where pulpwood was largely unutilised or unutilisable, royalty rates may be nominal so as to attract private industrial investment.

2) Where pulpwood utilising industries were established and were earning fair profits, the rate of royalty of forest produce may be such that the government can receive up to 50 per cent of the industrialists' profits.

3) In or near the forests where pulpwood utilising industries had reached an almost saturation point but entrepreneurs continued to compete for locating more industrial units, the

[54]It was also stated that royalty is only a token price primarily to assert the right of the government over such produce. This appeared to be a contradiction.

royalty rate of forest produce may be such that the governments share in profits could be in the range of 50 to 75 per cent.

These situations may be summarised in a tabular form as follows:

Description	*Price/royalty of the produce in the form of percentage share in industry's profits*	*Likely price/royalty rate per tonne of hardwoods assessed in 1971 (Rs.)*
Forests where the produce was not utilised	Nominal	Up to 5.00
Forests where processing industries had established sufficiently to utilise the produce	Less than 50 per cent	Up to 90.00
Forests where wood-based industries had established to a saturation point and keen competition existed to locate more processing units	More than 50 per cent	Up to 135.00

[Waheedkhan, 1971, p. 561]

The basis under reference did not appear to be logical and practicable even for determining royalty rate(s) of cellulosic materials, leave alone their prices.

1) It may be recalled that royalty has been defined as the land owner's (i) share in the natural product under consideration, or (ii) share in profits from the natural product. These two aspects cannot be equated. Conceptually, normal profits of a lessee of a natural resource such as forests or mineral deposits have to be a part of the cost of harnessing the natural produce. Even if the lessee did not undertake any processing of the produce, above normal profits or rent can accrue to him due to a positive difference between the average price received and the average cost incurred per unit of the natural produce. Price, in turn, would depend on a number of variables including place and time utilities added by the lessees. There could also be a possibility of losses in the process. In this sense where above normal profits were a net surplus or return to entrepreneurial ability, the government could not claim a share in them except through direct taxes.

2) The governments may like to reserve a share in the natural product not merely to assert their rights over the land but also, and more importantly, to meet the rights and privileges of the local people. This has been true of almost all formal agreements between the state governments and the paper manufacturing companies since 1940s.

3) The structure of the industry has substantially changed since the early 1970s when the basis under discussion was suggested. Significant differences with respect to locational advantages/disadvantages, size, age, technology used, extent of capacity utilisation, types and mix of paper and paperboards produced, etc. have emerged amongst the manufacturing units. Moreover, a majority of the paper and paperboards producers have adopted the pattern of multi-product firms/organisations. It would be extremely difficult to reliably monitor the total profits per unit of different types of outputs with paper and paperboard.

4) Profitability of integrated pulp and paper making units could change with the change in the mix of cellulosic materials. Thus, the suggested basis may retard the policy objective of encouraging enhanced use of unconventional materials in general and pulpwoods in particular.

5) The basis under discussion was suggested for fixing the royalty of pulpwoods. Its applicability to bamboo which has been and continues to be the single most important material for pulping was not discussed. This could possibly be due to the difference in the manner of supplying these two materials for the same purpose.[55] In that case, either the prevalent practice of bamboo supply to the pulp and paper making companies would have to be changed or one would have to think of a different basis for pricing bamboo.

6) There can be no definite principle for fixing the share of the cellulosic materials in the net profit earned by the paper and paperboard producers. There can also be an endless debate on whether the share of the forestry system/state governments should be in profits before or after taxes. Such issues could necessitate continued dependence on negotiations between the

[55]Mixed woods sold for pulping are harvested, sorted, and stacked by the FDs/FDCs whereas the paper mills carry out harvesting and all subsequent operations on bamboo from the leased natural forest areas.

concerned parties which, in turn, would mean continuation of the past practices. It may, however, be one of the considerations in fixing and revising the price (not royalty) of forest-based cellulosic materials.

7) Again, even after the forestry system's share in the net profits of the paper manufacturers is fixed, determination of prices of bamboo and hardwoods would be complicated by the fact that (i) the proportion of these materials in the furnish has varied and would continue to vary, and (ii) bamboo, though conventionally classified as a minor forest product, is one of the main products of natural forests while pulpwood has been a residual product.

8) The basis under discussion stipulates a direct relationship between the extent of development of forest-based industries in a catchment and the degree of competition amongst entrepreneurs to locate more of such industries in the same catchment on the one hand, and the share of pulpwood producers in the profits made by the industrial units on the other. This need not always be logical or desirable.

The existing industrial units can argue that the utility of the natural produce under consideration has been enhanced or that its supplies have become scarce due to their efforts at processing it, and they cannot, therefore, be penalised through raising the prices of the material in the form of raising the forestry system's share in their profits. The issue could be further complicated if additional investments in an industry in a specified area are detrimental to adequate utilisation of the capital already invested.

On the whole, however, competition amongst entrepreneurs could be desirable for accelerated pace of economic development in general and improved efficiency in utilisation of natural resources in particular. However, one can visualise that the newly established firms would find it difficult to earn even normal profits in the short run. Higher overheads including higher depreciation and interest costs could be some of the important reasons for that situation. That, in turn, could necessitate lower prices of cellulosic materials for the new units compared to the older ones. In fact, concessions in royalty rates of bamboo and hardwoods were given to new pulp and paper making units till recently.[56] The suggested principle, however, did not provide any room for such concessions. On the contrary, strict adherence to the principle would necessitate higher than the prevailing royalty rates/prices for the woody materials.[57]

This is not to say that new manufacturing units should necessarily be supplied with cellulosic materials at lower prices compared to the older units. On the contrary, it can be reasoned that the current administrative procedures and managerial practices can be substantially simplified if the extent of available incentives to the new industrial units can be appropriately specified/defined, and made available from one window.[58]

9) Under the present system, royalty has been charged on bamboo and hardwood *in situ*. Its conception as a cost of these materials can be very misleading, since the processing firms had to incur costs on harvesting, handling, transportation, and storage of bamboo and pulpwood (except harvesting costs for pulpwoods). These costs formed up to 90 per cent and 60 per cent of the total landed/mill yard costs of bamboo and mixed woods respectively, but varied significantly amongst paper mills. Therefore, any discussion of price of these materials as a share in the profits from the finished product(s) would necessitate replacement of the concept of royalty/price *in situ* by the concept of price at mill sites. One would have then to work backwards to account for handling and transport costs. Problems in adopting a standard procedure or a norm to handle this task have

[56]There was no clear or specific principle behind the concessional royalty rates of cellulosic raw materials supplied to the new pulp and paper making units. The periods of such concessional royalty rates varied between 5 to 10 years, and the rates varied from 33 per cent to 66 per cent of those applicable to the older manufacturing units.

[57]The authors of the suggestion themselves did not appear to be very sure of the consistency of their logic. Contrary to this reasoning for paper and paperboards making units, it was suggested that the newsprint industry in the country faced no competition worth the name and that industry should pay only a nominal royalty rate of Rs. 1-2 per tonne of pulpwood [Waheedkhan, 1971, p. 56].

[58]The practice of differential and favourable pricing of cellulosic materials for new paper and paperboard mills may have to be continued until the time the state governments and the GOI agree to simplify the existing procedures and practices.

been discussed earlier. A case by case solution may have to be worked out but that would not be easy.

The discussion in this sub-section may, thus, be concluded with the following observations:

—An attempt to find the price of cellulosic materials as a portion of profits earned by the pulp and paper producing companies would, most likely, create more problems than it would solve.

—The extent of profits of the paper and paperboards industry could be one of the considerations in determining the prices of bamboo and hardwoods used for pulping.

4.2.8 Price Equal to or Based on Market Price of Cellulosic Materials

An industry must function according to the market forces. This is relatively more true of the private sector but must be so for both the private and public sectors in the long run. A frequently repeated thought in the recent past, therefore, has been that if the paper industry: (i) could afford to pay for the fixed assets,[59] and other inputs like machinery, equipment, buildings, steel, cement, chemicals, electricity and coal, skilled manpower, semi-skilled and unskilled labour, etc. at the going market prices; and (ii) could sell 85 to 90 per cent of the output in the open market;[60] they should also be able and willing to pay for the cellulosic raw materials at their market prices. The reasoning is logical. It has also been relatively a more favoured thought with the forestry system due to (i) felt and perceived scarcities of the materials, and (ii) an unprecedented desire to enhance revenues from the forest produce.

The thought has also been shared by other organisations and individuals. As early as 1965, a Committee of the Planning Commission had observed that the royalty paid by the paper mills to the FDs was by no means a fair price, infact it was far below the commercial rate for bamboo. It would, therefore, be

[59] State Governments have been helping the manufacturing units in acquiring land at prices which may not equal the market prices.

[60] As noted earlier, whole of the output can now be sold at competitive prices.

misleading to use royalty as the basis for determining a reasonable market price for bamboo grown in plantations. [GOI, 1965, p. 25].[61]

Another important stage in this thought process appeared to be the Interim Report of the NCA, which observed that the supply of forest materials to forest-based industries was heavily subsidised [GOI, 1972, p. 65]. This statement has been quoted out of context by many people and organisations and widely differing policy implications have been derived from the term 'heavily subsidised'.

The term subsidy has, had different connotations at different points of time and in differing situations. At one stage it meant an aid granted by the House of Commons in England to the King for urgent needs of the kingdom whereas today in common parlance it often implies bribery. But the most popular use of the term subsidy has been to indicate a pecuniary assistance or aid or encouragement directly granted by a government to an individual or private commercial organisation deemed beneficial to the public. The financial assistance may be direct or indirect. There have, for instance, been cases of deliberately running the railways at a loss between the manufacturing centres and ports to reduce the cost of internal transport of goods with the objective of encouraging their sale in external markets. A more interesting example would be the subsidies offered by some governments during the Second World War on foodstuffs and clothing in order to keep down the cost of living so that the demand for higher wages could be checked. In essence, the supply of a good or service could be considered as subsidised if it is sold at a price lower than the cost of production.

In the context of this study, however, it must be recalled that the two most important historical facts behind nominal royalty rates of bamboo in India have been:

—The governments' desire to encourage the growth of

[61]The arguments appeared to be mixed. It was not clear whether the committee was making a case for fixing the royalty on bamboo on the basis of market price, or the thrust of their argument was that the prevailing royalty rates of bamboo could not be the basis for pricing the produce from man-made plantations.

bamboo-based paper industry in the country,[62] and

—Little or no costs incurred on managing the forests for producing bamboo which was considered to be a weed interfering with the growth of timber.

It can, therefore, be said that not too far back in the history of forest management in India, bamboo was considered either an almost free good or a good with negative value. Even in 1972 when the NCA presented its report, almost 100 per cent of the woody materials used by the industry emanated from natural forests. Data pertaining to production costs of these goods have been virtually non-existent, because these costs have been very low. Thus, the NCA could not have used the term 'subsidised' in the sense in which it is generally used.

This was also clear from the fact that almost in the same breath of stating that the supply of forest materials to forest-based industries was heavily subsidised, the NCA also stated that the *forester may be tempted to argue that he was supporting the industry* by charging *fantastically low prices for his bamboo as he could get a very high and profitable price in the open market. But our important pulp and paper industry would be adversely affected* if the forester was allowed to pick and choose his market for the bamboo [GOI, 1972, p. 65 emphasis added].[63] This should, however, not mean that the market price cannot be a consideration in determining the prices of woody materials used by the industry [64]

Returning to the main argument summarised at the beginning of this sub-section, it can also be said that a pricing policy which assessed both the forces of demand and supply, and sought to optimise net revenues or profits for the procedure(s) and net

[62]It could not be mere coincidence that the British Government in India in 1925 levied import duties on certain types of paper, and in 1932 the Bamboo Paper Industry (Protection) Act was promulgated. For details, see: Ahuja, 1980, p. 36.

[63]An interesting observation in this context has been that the society did tbe picking and chosing and not the forester. This could mean that the actions proposed or taken by the forestry system have been a mirror image of the socio-political dynamics.

[64]A further reading of their report indicated that the NCA's effort was to make a case for pricing the cellulosic materials of forest origin on the basis of the cost of raising them through man-made plantations. This aspect would be discussed in some detail in the next sub-section.

surplus to the consumers must weigh the market forces. This should generally be more relevant where the price(s) of the final product based on the good(s) under consideration was/were largely a function of the nature and the extent of the market, as was the case with paper and paperboard. At the same time, prices of industrial inputs such as bamboo and woods cannot be equal to their market prices and/or prices in competing uses for a number of reasons:

1) The demand (not requirements) of woody materials and the demand for paper and paperboards at various price levels was not known. Similarly, costs of production to meet different levels of demand were also not known. These problems would be the starting point of a number of conceptual and practical problems in determining the market prices (Annexure 1).

2) The grade specifications of the goods in the seemingly competitive uses could be, in fact, have been significantly different. For instance, species and grades of bamboo used by households, cottage industries, construction industry, etc. vary not only amongst the activities but also within them. More importantly, bamboo used by the pulp and paper industry was termed as *kachra* or non-marketable. In other words, bamboo in every form, including dry/green/broken/crooked/diseased was usable for the purpose.[65] A satisfactory basis to compare the weights of industrial and commercial bamboo remained to be found.

The situation with regard to pulpwoods may not be so clearly differentiable. In the past, the pulp and paper industry was extremely choosy about pulp woods from natural forests, but that has also been true with at least one other set of buyers: the charcoal makers. The situation has changed due to improvements in pulping technology, and the industry can be encouraged to use almost any size or type of pulpwoods.

3) The real values of both bamboo and pulpwoods differ substantially amongst the so-called competing uses which vary from manufacture of sophisticated decorative goods to meeting the basic necessities of life. A natural consequence is that there are various types of markets for seemingly similar goods and

[65]A considerable portion of bamboo used for pulping may belong to commercial category, but the industry has been using it mainly because sorting the produce has not been financially and managerially feasible.

their prices vary substantially due to variations in time, place, and form.

The issue is further complicated by the prolongation of the *nistar* rights which, once again, varied not only amongst states but also within the regions/districts of the states. In one state, for example, choicest bamboo was supplied at people's doorsteps residing near the forests at Rs. 55 to Rs. 65 per tonne while comparable quality of the material fetched around Rs. 1,500 per tonne in wholesale and Rs. 2,500 in retail urban markets. Similarly, in one district of another state, 5,00,000 whole bamboos were brought in annually by the FD covering a distance of over 1,000 km and supplied to the local people at the *nistar* rates.[66]

It must also be noted here that the extent of the real benefits derived by the people/families who avail of *nistar* rights have not been studied. Data on financial and manpower costs to the FDs were also not available. Moreover, little thought had been given to the period up to which such rights may be continued though it has been argued that these would have to be progressively abolished [Kothari, 1971].

4) The prices of a good payable by a bulk consumer and numerous small consumers have never been the same. In most cases, such prices have not even been comparable. Moreover, the paper and paperboard manufacturers have been managing all the operations on the portion of the produce used by them while such functions must be performed by the FDs for the other users.

In some states the agreements between the governments and the pulp and paper making companies necessitated that the latter must hand over to the former specified quantities of *marketable*, i.e., selected bamboo out of the total extracted by them. This saved considerable administrative and management efforts of the FDs.

5) The intrinsic worth and thus the individual and societal values of a good differ amongst its uses. One can argue that the social values of wood used for cooking or bamboo used for

[66]This right was introduced when bamboo was grown in the natural forests in the district. Its continuation could even be interpreted as a reward to the local people in return for their contribution (which may be very small) in depriving the approachable/local forest areas of bamboo.

hutments and/or low income group housing were higher than their value for pulping. The argument may not be refuted even if paper and paperboards were products with multiple values.

On the other hand, it could be recalled that (i) a substantial portion of forest outputs were supplied to or availed of by the people either free (except the opportunity cost of the time used in collecting them) or at very nominal prices; (ii) the industry consumed only five per cent of the forestry system's outputs and this could be reduced to less than two per cent if two-thirds of the forest land in India was managed to yield the lowest level of its estimated potential outputs (Chapter 3); and (iii) the *indirect* taxes paid to the central government and the state governments by some of the large sized integrated pulp and paper mills averaged over Rs. 900 per tonne of bamboo used by them.[67]

This is not to argue that industrial uses of the material should be construed as socially and economically more desirable. The reasoning is that (i) a basis for assessing the relative societal values of the goods in these uses must be found before one can logically use market price as one of the bases for pricing them, and (ii) that the issue can be objectively debated in conjunction with an assessment of the forestry system's potential to meet different types of demands for woody materials.

6) The industry has argued that they have themselves contributed to the growth and extension of the markets for cellulosic materials, and this was particularly true for bamboo. The reasoning may be valid only partially as the demand for bamboo also increased due to the disproportionate rise in timber prices. At the same time, and even at the cost of repetition, it can be recalled that as late as 1940s bamboo had little economic use. It was considered a wild grass and the policy makers and foresters were keen to invite the users so that whatever they could receive would be a net surplus. This also seemed to be the reason for the use of the term royalty of bamboo and for adhocism in fixing royalty rates.

7) The market prices were ***not representative*** in the economic sense of the term. Only a small portion of the industrial

[67]The average per tonne of cellulosic materials would be lower, but the relevant data were not available.

bamboo reached the market. A good portion of the marketable produce is also reserved to protect the rights and privileges of the local people. More importantly, the behaviour of the market price of bamboo could not be predicted if the entire output was placed for open sale.

4.2.8.1 *Use of Market Price of Bamboo and Hardwoods as a Basis of Their Pricing for Industrial Uses: Two Examples*

Irrespective of the logic of the preceding arguments, at least one state government has decided to price bamboo supplied to the paper and paperboard manufacturers on the basis of its market prices. In 1982 the state legislature promulgated an Act to empower the government to modify the existing agreements with the purchasers of forest materials, to remove any ceiling on the increase in the prices at the time of their revision subject to a maximum of the market price of the produce under consideration. The market price was supposed to be determined by the state government in a prescribed and *standard* manner.

Market price is a very ticklish concept and raises a whole lot of issues such as types of relevant markets, extent of markets, types and quality of the produce under reference, use(s) of the produce, and the relevent period. It may, therefore, be difficult for a state government to prescribe the stipulated *standard* procedures for determining the market price particularly for a good which is not homogeneous.

The state government was probably conscious of such problems while proceeding to determine the market price of bamboo. A set of draft rules was prepared for this purpose and suggestions/objections were invited from persons and parties likely to be affected by them. However, there was no response from any person or party within the stipulated period of one month. This resulted in the announcement/publication of the "Supply of Forest Produce by Government (Revision of Agreements) Rules" in June 1983.

The rules provided that the government was obliged to explain the basis of determining the market price (value) of the forest produce under consideration through a note. This note was to be sent to the purchaser(s) along with the notice for revision(s) in the existing agreement(s) and the process of determining the market price was to consider the following aspects:

—The prices obtained in the open and negotiated sales of such produce effected by the government or the FD or the FDC in the state, during a period of 12 months preceding six months prior to the date of commencement of the supply year. But, if such sales did not take place during the stated period or if in the opinion of the government the number of sales and the quantity of the produce transacted were not adequate, the price(s) obtained in such sales during a period of 24 months preceding the period of six months prior to the date of commencement of the supply year could be considered; and such sales could even be ignored if the government thought that the quantity of forest produce involved was not sufficiently large to serve as *representative* for the purpose; but if there were more than one sales during the chosen period, weights would be assigned to the quantities sold.

—The general trend in prices of the forest produce during the relevant period.

—The differences in distance between the areas where the sales (referred to earlier) had taken place and the purchaser's location to which weights must be assigned.

—Open and negotiated sales of such produce in one or more of the adjoining states if no sales took place within the state during the relevant period(s).

—Any other factor considered relevant by the government for the purpose.

This opened a number of issues. For instance, weights were to be assigned to the quantities sold in more than one sale during the chosen period, but it must be extremely difficult to:

—Specify the basis of such weights acceptable to all the concerned parties;

—Logically determine the weights where the quantities of the produce were significantly different in different sales such that there could be a reasonable ground to differentiate between the financial strength and/or the risk bearing capacities of the buyers;

—Determine the weights only on the basis of the quantities sold at different places not only with respect to distance from the users' locations but also with respect to variables such as accessibility to road/railroad heads.

The state government and the FD were not fully willing to share their thoughts, or the data, or the documents on such issues. However, in exercise of the powers derived from the 1982 Act and the 1983 Rules, the state government proposed to raise the royalty rate of bamboo to Rs. 230 per ADT starting from October 1983. The proposed rate was considered to be lower than the market price of industrial bamboo during the preceding year. A note explaining the basis for arriving at the market price during the supply year 1982-83 was also provided to a paper and paperboard manufacturing company which had been buying bamboo for over 35 years. This has been summarised here for an understanding of the nature and type of problems faced by the state government in arriving at the proposed price.

The state government recognised that the price obtained in the open and negotiated sales from April 1982 to March 1983 should form the basis for determining the market price of bamboo. During that period, however, only outright sales of flowered bamboo coupes had taken place in one of the forest divisions in the state. An average price of Rs. 424 per MT was obtained.[68] But, this price was computed from estimated and not actual outputs and could, therefore, not represent the actual price. Moreover, these sales of bamboo were for commercial and not industrial use(s). This necessitated the use of data pertaining to open and negotiated sales of bamboo for a period of 24 months during April 1981 to March 1983 as there had been:

i) A negotiated sale of 13,000 NT of industrial bamboo from one of the forest ranges in the state to a paper making company in a neighbouring state during June 1981 to March 1983 at a royalty rate of Rs. 195 per NT which worked out to be

[68]Some forest officers opined that pricing flowered bamboo has been a hit and miss affair. They also opined that it could be priced on the basis of a fixed percentage of the normal bamboo, but there has been little objective thinking on the subject.

Rs. 264 per ADT on bundle-to-a-tonne basis considering 67.7 bundles per one tonne.

ii) Open sales of nearly 19,700 NT of bamboo from three forest divisions in the state in October 1981 at the average price of a little over Rs. 340 per NT or Rs. 490 per ADT on bundle-to-a-tonne ratio basis where 72 bundles were considered as one NT.

Once again, the latter could not be relevant for fixing the market price of industrial bamboo because the buyers harvested only full length bamboos. In fact, that appeared to be one of the main reasons for the high price.

The government reasoned that the first case pertained to a relatively inaccessible forest area compared to the areas from where bamboo was to be supplied to the company and it would, therefore, be appropriate to assume that the market price of comparable bamboo from the latter would be considerably higher than the price received from the former. It was also reasoned that since the sales pertained to the supply year 1980-81, it was necessary to consider the rise in the prices of all types of forest produce during the intervening period for determining the price of bamboo starting from 1982-83. This, however, could not be possible because an adequate number of open and negotiated sales of industrial bamboo had not taken place. The note, thus, concluded that considering:

—the price of Rs. 264 per ADT of industrial bamboo received from a comparatively inaccessible area in 1980-81,

—the average price of Rs. 490 per ADT for a mixture of commercial and industrial bamboo realised in 1980-81, and

—the general rise in prices of forest produce since 1980-81;

it would be reasonable to *determine* (not estimate) the market price of industrial bamboo at Rs. 300 per MT, and the proposed royalty of Rs. 230 per ADT was significantly lower than the market price of industrial bamboo. The issue remained unsettled at least till the end of 1987. The main question has been whether the market price estimated by the government was representative and applicable to the company. We could not get the details of the reasoning from either side. A few aspects which may be relevant to discussions between the government and the company could, however, be thought of:

1) The market price determined by the government was primarily based on the sale of 13,000 NT or 9,600 ADT of bamboo to a paper making company in another state over a period of two years. On the average, this did not form a significant portion of the company's total annual consumption of cellulosic materials. During that period, the company paid Rs. 140 per ADT for the bamboo harvested from the forest areas in the same state as the company's mills location. For the bamboo purchased by them from the (other) state under reference, the additional payment amounted to Rs. 1.2 million over the two-year period. This may appear to be a large amount in absolute terms but it accounted for a very small portion of their normal annual cellulosic raw material costs such that the impact on costs per unit of their final outputs could have been insignificant or even unnoticeable.

In essence, it needs to be known whether the quantities of bamboo supplied to a paper making company with which a government does not have regular arrangement, are marginal or substantial vis-a-vis the company's total demand. The answer would indicate whether or not the prices paid can have a noticeable impact on the cost of the final output(s) and the financial health of the company.

2) It can be hypothesised that the bamboo purchased by a company from another state with which it did not have regular arrangements was used to run its paper mills at higher than the rated capacity such that the fixed costs could be distributed over a larger quantity of final outputs. Financial, economic, and managerial implications of such situations can be positive and may easily be understood by all concerned.

3) It could also be hypothesised that a paper making company located in another state may be strong competitor of another company who has a regular arrangement for supply of cellulosic materials with the government of the state of their location. If that be the case, the former could or may purchase large quantities of bamboo at higher prices for a year or two from the other state with the objectives of starving the competitor of cellulosic raw materials, or for gaining market leadership such that they may be able to purchase the raw materials or sell the final outputs largely on their own terms in the future.

Unmindful of these and such other problems, it has even been suggested that price of woody materials supplied to forest-based industries should be based on the current market prices rather than on the prices of the preceding year or years. That would supposedly enable the forestry system to take full advantage of rising prices of the produce. The questions which remain unanswered pertain to administrative and managerial feasibilities, and social desirability of such a step.

In another case in the same state, market price of hardwoods was accepted as a relevant consideration for fixing the royalty of the material payable by a pulp and paper making company in the middle of 1970s. At that time, only the fuelwood and charcoal merchants were the large scale buyers of the material and the price realised by the forestry system varied from Rs. 4 to Rs. 13 with an average of Rs. 8 to 9 per standard stack of 2 cu m (r). The FDC in the state could also face the problem of selling large quantities of hardwoods which emanated from clear-felling of the forest areas.

For the paper making company, the royalty/price *in situ* was fixed at Rs. 16 per standard stack. This was higher than the market price but was agreed upon by the parties. The logic was that the market price depended on and varied widely amongst the forest coupes for several reasons including:

- —Location of the material vis-a-vis the location of the bulk buyers,
- —Variations in solid contents of fuelwood stacks,
- —Proportion in small and large sized billets in fuelwood stacks,
- —State of the charcoal market, and
- —Time of the year (winter/summer) when the material was put to sale, etc.

Though the royalty rate of bamboo was lower, yet the landed cost of the hardwoods were to be lower because large quantities were available in a concentrated area close to the company's mills.

This shows that the market price of the material was successfully used for initiating the negotiations for fixing the royalty rate/price, and modifications in the market price could be made to the advantage of both the parties. This could not be possible without an environment of mutual trust, or if either side insisted that market price of the material should also be the price for its industrial use. The objectivity has weakened over time, and there were considerable difficulties in revising the price of pulpwoods effective from the late 1970s or early 1980s in this case.

The two cases and the preceding discussion in this sub-section should suffice to conclude that (i) the economic relationship between the forestry system and the users of cellulosic materials needs to be redefined in the context of the changed circumstance, and (ii) the market prices of cellulosic materials can be one of the several considerations in deciding upon a sustainable pricing policy for them.

4.2.9 Prices Based on Cost of Production of Cellulosic Materials

Producers or suppliers of a good normally expect the good to sell at a price equal to or higher than the cost of production. Application of this logic for the forest produce was, probably, made in 1965 by a committee of the Planning Commission which argued that the prevailing royalty rates of bamboo could not be a basis for pricing the produce from man-made plantations [GOI, 1965, p. 25]. The committee then proceeded to point out that the delivered costs of bamboo at the paper mills was at least Rs. 100 per tonne of which the costs of felling and carting to road sides were Rs. 30 per tonne. The lower the costs of transport and handling, the higher the royalty (or price) the mills could pay. It was, therefore, argued that if :

—bamboo could be grown in plantations so as to obtain concentrated supplies in easily accessible sites near the mills, a royalty of Rs. 30 per tonne would not be considered too high, and

—bamboo in plantations could be grown at a cost so as to leave a fair margin of profit when sold at that (Rs. 30 per tonne) rate, suitable forest land should be utilised for the purpose to augment the supplies of long fibred material in the plains [GOI, 1965, pp. 25-26].

The committee's reasoning does not appear to be clear. On the one hand, a case is made for higher royalty rates without enhancing the landed costs of the material at mill yards, while on the other, the argument is to allocate suitable land in the plains for bamboo plantations to enable a reasonable return on the investment with given price of the output.

The committee, however, presented a methodology to work out the compound interest rate at which capital invested in raising the plantations, including a reasonable rent on land, could result in a normal forest so as to yield bamboo in perpetuity.[69] For want of data on the actual costs of the activity, hypothetical data were used to simulate three bamboo plantation projects with different fixed and variable costs per hectare, and changing prices of the outputs. In each of these cases, the project cycle was assumed to be 32 years with the first crop in the eighth year and six subsequent crops every fourth year. The yields were estimated at 10 tonnes per hectare per crop except 15 tonnes per hectare in the terminal year. The land expectation value, i.e., the land rent was assumed to be zero. Rates of return on investment at varying prices of bamboo were estimated [GOI, 1965, pp. 26-29].

This methodology cannot be used to argue a case for pricing the outputs on the basis of the production costs. The issue of the rate of return on investment which the project authority/ FDs/FDCs may wish to earn or the basis for deciding the rate of return on investment was also left open. Two important contributions of the committee's work could, however, be in terms of :

1) Visualising and demonstrating the impact of the changes in the rate of interest on a break-even price per unit of the output and vice versa, and

2) Making a case, if necessary, for financial help or subsidy from the governments to the FDs/FDCs to enable them to plant derelict areas at reasonable costs and to earn a reasonable rate of interest on the investments [GOI, 1965, pp. 28-29].

The same committee, however, also demonstrated that the simplest method of arriving at the minimum rate of royalty (price) chargeable for pulpwood grown in plantations would be

[69]The argument appeared to be that adequate capital can be invested only if the rate of return is adequate.

to compound all costs incurred on creating and tending a plantation till it is ready for felling and then work out the cost per tonne of the output obtained. In a model to demonstrate the use of this method, it was assumed that (i) the cost of formation would be Rs. *C* per hectare, (ii) the annual maintenance costs, including interest on loan would be Rs. *e* per hectare, (iii) gestation period of the plantation would be *r* years, (iv) the yearly rate of compound interest on the loan raised for the plantation would be *p* per cent,[70] (v) yield per hectare on maturity of the plantation would be *y* tonnes, and (vi) cost of production per tonne of wood would be Rs. *R*.

Substituting X for 1.op,[71] the capitalised value of *C* at the end of *r* years = CX^r, and
the capitalised value of annual expenses would be

$$e + eX + eX^2 \ldots\ldots\ldots + eX^{r-1}$$

$$= e\,\frac{(X^r - 1)^{72}}{(X - 1)}$$

Therefore, the capitalised value of total costs would be

$$\frac{(X^r + e\,(X^r - 1)}{X - 1}, \text{ and}$$

The cost of production per tonne of pulpwood would be

$$\frac{\dfrac{CX^r + e\,(X^r - 1)}{X - 1}}{Y} = R$$

Assuming further that
C = Rs. 750
e = 25/40/50
p = 4, 5, 6,......10 per cent
y = 70 tonnes
r = 10 years

Costs per tonne of the material (Rupees) were worked out as follows:

e(Rs)	*p (percentages)*				
	4	*5*	*6*	*8*	*10*
25	20.14	21.95	32.90	28.31	33.48
40	22.71	24.65	26.73	31.42	36.88
50	24.43	26.45	28.62	33.49	39.16

[GOI, 1965, pp. 29-30]

[70] The use of the word compound was unnecessary and even confusing.

[71] Should have been written as 1+P.

[72] Assumed that money for this purpose was borrowed at the *end* of each year.

The exercise was of interest as it demonstrated that

—changes in annual operation and maintenance costs,

—changes in discount rate, etc.

can make a considerable difference in the capitalised production cost of an economic good. At the same time, there were a number of conceptual and practical problems with it.

1) It was assumed that the money required to meet the O+M costs was borrowed at the end of the year. On the contrary, an accepted convention for financial and economic analyses/appraisal of development projects is to assume that all costs are incurred in the beginning of the year.

2) It was stated that all costs including land rent must be considered, but the example under reference made no mention of the rent of forest land. In the earlier case it was assumed to be zero. This may serve to strengthen the arguments presented earlier that determination of rent of forest land would be extremely difficult if not impossible.[73]

3) Choice of a compound or discount rate has been one of the major problems in assessing financial and/or economic feasibility of developmental projects/activities. The committee could have at least indicated an approach to solve this problem, particularly when one of the stated objectives of raising plantations tree crops was to ensure adequate returns on the investments made by the forestry system.

4) It could be surprising that the expert committee which consisted of very eminent pioneers of the Indian Forest Service did not consider it necessary to (i) assess the outputs from, and (ii) the annual operational and maintenance costs of eucalyptus plantations. This was necessary for technical as well as financial reasons.

One could argue that the activity may not continue beyond the first (seedling) crop. That would be conceptually valid. In reality, however, the argument cannot hold good because (i) land under reference had to be used for tree crops, (ii) the establishment costs of the coppice crops were negligible or zero,

[73]This may also show that a number of suggestions have been continuously made by responsible individuals and organisations without any efforts at assessing their feasibility or documenting the problems in that process.

(iii) the operation and maintenance costs of the coppice crops could be lower than those for the main crop, (iv) the maturity/rotation age for the coppice crops could be shorter than for the seedling crop, (v) the man-made tree plantations were/are supposedly raised to meet the industrial demand for woody raw materials which continue to be in short supply, etc. In spite of these considerations, one may choose not to account for the returns from or costs of the coppice crops. In that case, however, comparisons of the annual costs and returns would necessitate estimation of salvage value in the terminal year of the project.

The next main stages were the interim and final reports of the NCA which appeared in 1972 and 1976. Thc Commission suggested that for supplying cellulosic raw materials to the industry the forester must be *guaranteed remunerative prices* which would pay for the *cost of raising* the materials through man-made plantations, including costs of clearfelling and maintenance of production forests and also leave a profit [GOI, 1972, p. 67 ; GOI, 1976c, p. 27, emphasis added].

At the same time, the NCA also pointed to the need to *strike a mean such that organised bamboo plantations* were used *partly to supply the industries at open market prices* depending on industrial growth and the *capacity of the forester to expand his plantations* [GOI, 1972, p. 67 emphasis added]. But, it has been discussed earlier that the NCA did not favour market process as the basis for pricing the supplies to the pulp and paper industry. It can, therefore, be said that the NCA's reasoning on the subject has been vague. Moreover, the statements summarised here could mean that the Commission's suggestions would have been met even if the price of cellulosic materials supplied to the industry was somewhat lower than the cost of raising them through man-made plantations, and that situation could be justified on the grounds of economic, social, cultural, and administrative importance of paper and paperboards. That, however, did not appear to be the case.

On the contrary, the Commission attempted to justify a remunerative price for the materials which, to repeat, could pay for the entire cost of production and leave a profit for the FDs/FDCs [GOI, 1972, p. 67]. It was further argued that the industry should be able to pay still higher prices as the man-made plantations can be set up within a reasonable distance from the processing units, thus resulting in considerable savings on harvesting and transportation costs compared to those from natural forests. A hope was also expressed that it should be possible for both the parties to take a reasonable view to make man-made *plantations a profitable venture as these could not be sustained on subsidy* [GOI, 1972, pp. 66-67 ; GOI, 1976c, p. 73 emphasis added].

It would, therefore, be difficult to conclusively think whether the NCA was suggesting a logical and sustainable basis for pricing the materials used by forest-based industries or it was presenting a case for raising man-made plantations. Though, man-made plantations have not been as popular or successful as they were expected to be in the 1970s, yet, it cannot suffice to say that man-made plantations should not be raised, and that cost of production should not be a variable in pricing the materials used by the forest-based industries. The merits and demerits/problems of practising the principle must, therefore, be assessed. This may be better done in the context of a real life experience summarised in the next sub-section.

4.2.9.1 Determination of Stumpage Price of Plantation Grown Eucalyptus : A Case

Though the practice of single species tree plantations by the FDs in India is almost as old as the art and science of forest management in the country, the activity has been carried out on a considerable scale during the last three decades or so. The total area planted in this manner could be in the neighbourhood of three million hectares. The main thrust appears to have been on eucalyptus followed by teak, sissoo, casurina, bamboo, etc. Data on the area under each species, financial costs incurred, rates of growth, rotation lengths and yields, were not available for many situations in the country. A few experiences have, however, been documented. One of these pertains to large scale eucalyptus plantations in the Tarai and Bhabhar Forest Divisions of Uttar Pradesh which was started in 1962 [Misra, 1980a].

The plantations were done on sparsely wooded but productive forest land. The main objective was to produce raw materials for the pulp and paper industry of the state. It has been claimed as one of the most successful examples of man-made plantations in the country in spite of the fact that the observed yield was around 30 tonnes per hectare per rotation against the initial estimate of 70 tonnes. At the same time, the crop matured faster than expected and the rotation was reduced from the initial estimate of 10 years to 8 years [Misra, 1980a, pp. 1-2].

The plantation started in 1962 should have been harvested in 1970-71, but that was not done. An exercise to decide on the policy and principles for fixing the stumpage price for the produce from these plantations was initiated in 1972. Some of the identified bases were :

—Prices obtained at small auction sales of eucalyptus wood in various forest divisions in the state;
—Sales prices of eucalyptus wood obtained in other states;
—Landed prices of eucalyptus wood for pulping as recommended by the NCA in 1972;
—Landed prices of wood arrived at by the World Bank in their appraisal of the eucalyptus plantations project in the state; and
—Internal rate of return [Misra, 1980a, pp. 2-3].

Senior forest officers in the state discussed these approaches and decided that an adequate rate of return on the investment would be the most relevant and reliable basis for determining the price. It was also suggested that a scenario of stumpage prices be used and corresponding IRRs be computed to be able to decide on an adequate IRR. Once an adequate IRR was decided upon, the corresponding stumpage price would be charged on the user industrial unit(s).[74]

Data on actual annual costs were available. But, yields from the first and three subsequent coppice crops had to be estimated. Though the management system accepted by the department was that the plantations would yield four harvests at intervals of eight years each,[75] yet any stage of the crop could be logically used as a cut-off point to compute the IRRs. That, how-

[74]The cause and effect relationship was not clear.

[75]Rotations for the seedling (first) and the coppice crops was taken to be of the same duration.

ever, would have necessitated an estimation of the value of the standing crop at that point in time. A relatively simpler alternative chosen in this case was to follow the activity through its total cycle [Misra, 1980a, p. 4].

The major problem was that a 100-150 TPD integrated pulp and paper mill which was expected to be established by the early 1970s in this catchment did not come up, and an existing pulp and paper making unit was willing to buy only 75,000 tonnes of the eucalyptus wood per annum. This resulted in a perception that the produce from the entire mature crop may either not be sold or the prices may fall below the financially viable level. The Departments of Forests and Industries, therefore, jointly decided that even if more than 73,000 hectares, i.e., the area planted during 1962-65 were due or over due for harvesting and additional areas were to mature every year, annual harvesting should be limited to 2,500 hectares (i.e. 75,000 tonnes), and the rest of the mature crop should be held back till 1978 when the anticipated pulp and paper mill could be operative.

This also necessitated an adjustment in the harvesting pattern starting from 1978 to meet the anticipated paper mill's demand for eucalyptus wood and to avoid wide annual fluctuations in its supplies. The result was to fix the harvesting pattern to satbilise the supplies to around 4,50,000 tonnes per annum.[76] The next step was to compute the IRRs and the consequent stumpage price on the basis of the proposed harvesting pattern (not summarised here). But the stumpage price thus worked out was not acceptable to the paper making company which wanted to buy 75,000 tonnes wood per annum. Their argument was that the decision to withhold mature crops for supply to the proposed paper mills resulted in higher stumpage price for any IRR and thus amounted to a penalty on them. The existing company, therefore, proposed that the entire plantation area be divided into two parts: one pledged for supply of raw materials to them and the other to the proposed mill, and the stumpage prices be computed separately for each

[76]Theoretically the supply from a plantation area should go to zero on harvesting the third coppice crop, but the area could be replanted to maintain the flow of output.

This proposal, however, was not acceptable to the FD for two reasons. First, delineation of the planted area in two segments was considered to be not feasible. Second, and more important, the FD considered it inappropriate to sell the same material at different prices to the existing and the proposed paper making unit.[77] But the issue had to be re-examined.

It may be recalled that the decision to withhold mature stocks was taken for two reasons: (i) to conserve the raw material for supplying to a proposed paper mill in the state, and (ii) to avoid an apprehended fall in the market price of the produce. Experiences in other states, however, indicated that eucalyptus wood had a ready market and fetched good price. It was also noted that withholding the trees beyond the rotation age resulted in a hard brown core in the wood which made it less suitable for pulping. It was, therefore, decided that the whole of the mature crop be harvested and the material may be offered for sale to paper mills in other states also [Misra, 1980a, p. 5].

The felling schedule proposed earlier was, thus, revised. The decision was taken to harvest the areas planted during 1962 to 1966 as fast as possible, and thereafter to fell each plantation area as it matured. At the same time, it was assumed that the coppice crops would be harvested in the 9th, 18th, 27th and 34th years instead of 8th, 16th, 24th and 32nd years of plantation. Columns 1-4 of Table 4.1 present the relevant data. The Table also presents the cost and output streams.[78]

An important observation from Table 4.1 would be that even after ignoring the years 1974-78 cyclic rise and fall was expected in annual outputs from 2,83,000 tonnes to 5,41,000 tonnes. This situation could have been avoided if the plantation activity was evenly spread, and the crops were harvested in the years of their maturity.

For computing the IRRs, the FD compounded the cost stream for the years 1962—1973 to 1974 while both cost and

[77]This meant that the price payable by the proposed industrial unit had to be higher while quality of the material could be relatively inferior due to overaging.

[78]Costs for the years 1962 to 1973 are actuals while those for the period 1974 to 1978 and 1979 to 1981 were estimated at the rate of Rs. 1,025 per hectare and Rs. 1,230 per hectare, respectively, including all fixed and variable costs [Misra, 1980a, various pages].

TABLE 4.1 : **Annual costs of and outputs from 1,24,180 hectare of eucalyptus plantations raised during 1962-1981 in Uttar Pradesh**

Costs : Rs. 000
Output : 000 tonnes

Years	*Area planted (hectares)*	*Annual costs*	*Annual outputs*	*Costs in 1974 discounted at 14 %*	*Outputs in 1974 discounted at 14%*
(*1*)	(*2*)	(*3*)	(*4*)	(*5*)	(*6*)
1 = 1962	419	378	—	1821	—
2 = 1963	285	395	—	1669	—
3 = 1964	2549	1568	—	5813	—
4 = 1965	4101	2345	—	7626	—
5 = 1966	5578	3945	—	11255	—
6 = 1967	6013	3268	—	8177	—
7 = 1968	6530	3501	—	7685	—
8 = 1969	6531	5367	—	10332	—
9 = 1970	7033	4539	—	7666	—
10 = 1971	5266	4568	—	6765	—
11 = 1972	7926	5864	—	7617	—
12 = 1973	7890	6349	—	7238	—
13 = 1974	8000	8200	0.30	8200	30
14 = 1975	8000	8603	2.00	7545	175
15 = 1976	8000	10887	2.00	8372	154
16 = 1977	8000	10887	2.00	7349	135

17 = 1978	8000	10887	2.00	6445	118
18 = 1979	8000	12487	2.83	6481	147
19 = 1980	8000	14959	2.83	6548	129
20 = 1981	8000	14359	2.83	5744	113
21 = 1982	—	4559	2.83	1600	99
22 = 1983	—	4559	2.61	1404	80
23 = 1984	—	4195	4.41	1133	119
24 = 1985	—	6934	4.41	1443	104
25 = 1986	—	6934	4.41	1442	92
26 = 1987	—	6934	4.41	1262	80
27 = 1988	—	6934	5.28	1109	85
28 = 1989	—	8271	5.28	1158	74
29 = 1990	—	8271	2.98	1017	37
30 = 1991	—	4559	2.98	492	32
31 = 1992	—	4559	2.74	433	26
32 = 1993	—	4195	4.54	348	38
33 = 1994	—	6934	4.54	506	33
34 = 1995	—	6934	4.54	444	29
35 = 1996	—	6934	4.54	388	25
36 = 1997	—	6934	5.41	340	27
37 = 1998	—	8271	5.41	356	23
38 = 1999	—	8271	2.98	314	11
39 = 2000	—	4559	2.98	150	10
40 = 2001	—	4559	2.74	132	8
41 = 2002	—	3712	4.54	97	12
42 = 2003	—	3712	4.54	82	10
43 = 2004	—	3712	4.54	74	9
44 = 2005	—	3712	4.54	63	8

TABLE 4.1 (*Contd.*)

(*1*)	(2)	(*3*)	(*4*)	(*5*)	(*6*)
45 = 2006	—	3712	5.41	56	8
46 = 2007	—	3712	5.41	48	7
47 = 2008	—	3712	2.98	45	4
48 = 2009	—	456	2.98	5	3
49 = 2010	—	371	2.43	3	2
50 = 2011	—	371	2.43	3	2
51 = 2012	—	371	2.43	3	2
52 = 2013	—	371	2.43	2	2
53 = 2014	—	371	2.43	2	1
54 = 2015	—	371	2.43	2	1
55 = 2016	—	371	2.43	2	1
Total				156305	2105

Price per tonne output=Rs. 74.25.

Source : Misra, D.N., 1980 a, pp. 10-19.

return streams starting from 1975 were discounted to 1974, i.e., the first year of harvesting the crop. The exercise was done at varying discount rates. Corresponding stumpage prices to equalise the cost and return streams were computed. Since the discount rate and/or the stumpage price finally chosen were 14 per cent and Rs. 74.25 per tonne, respectively, of eucalyptus wood, Table 4.1 demonstrates only these computations.

4.2.9.2 *Some Methodological and Practical Issues in Using Cost of Production as the Basis for Pricing Woody Materials*

This case, a perusal of related literature, and discussions with a cross-section of people brought out a number of facts and issues.

1) One of the main concerns of the managers and administrators of plantation projects till about a decade back was to ensure an adequate price for the output in the open market to be able to earn a reasonable return on the investment. Thus, the perceptions of absolute scarcity of woody materials were a matter of recent past and could at least partly be a result of inadequate success of some major forestry development programmes in general and man-made plantations projects in particular.

2) In spite of the fact that raising single species tree plantations to meet the raw material requirements of forest-based industries has been a well established practice and the foresters were possessed with the requisite/technical capabilities, the actual costs have been substantially higher and/or observed outputs have been lower than those estimated at the time of formulating the plantations projects. These, in turn, can be the most important bottleneck in practising the principle of cost-based pricing. To explain this point, we carried out an exercise on the lines of the case discussed in the preceding sub-section. It assumed that :

—A eucalyptus plantation was to be raised to specially feed an integrated pulp and paper making unit with a capacity of 30,000 TPA.

—Of the total annual requirements of cellulosic materials,

the pulp mill could use 70 per cent of eucalyptus wood= 55,000 tonnes per annum.

—Eucalyptus plantations could easily generate 70 tonnes of dry wood per hectare per rotation of eight years[79] so that the paper mills' demand for eucalyptus wood can be met if plantations on 785 (say 800) hectares are ready for harvesting every year.

—The total land to be placed under eucalyptus for assured supply of the material should be 6,400 hectares and this area should be planted @ 800 hectares per year over an eight year period.

—The plantations costs in the first year would be comparable to those estimated for the year 1981 in the case discussed earlier (Table 4.1) but the operation and management costs would be substantially higher.

—The project authority would like to earn 14 per cent per annum on the investments made in the plantations.

The exercise has been presented in Table 4.2. A perusal of the Table can show that the breakeven stumpage price including normal returns for the project authority would be Rs. 32.85 =Rs. 33 per tonne as against Rs. 74 per tonne in the previous case. The substantial difference has been caused not only by the difference in yield levels but also, and more importantly, by timely harvesting of the mature crops.

This could also indicate that the use of the principle of cost-based pricing would necessitate a price determination exercise for each situation/paper making company. Let us assume that a reasonably uniform figure in respect of plantation costs per hectare of selected tree/bamboo species can be adopted.[80] But

[79]This was assumed for the plantations project discussed in the preceding sub-section. The actual yield was 30 tonnes per hectare per rotation but reasons for the difference between expected and actual yields have not been documented. On the contrary, there have been examples of eucalyptus plantations with yields varying between 80 to 150 tonnes per rotation. Availability of irrigation could reduce the rotation cycle or further enhance the yield.

[80]This can be possible if the governments agree to compensate/subsidise the FDs/FDCs for the difference in costs of planting relatively more difficult or degraded areas compared with the easily accessible and/or productive forest land. The policy would serve the objective of improved management and conservation of the most scarce of the productive resources.

TABLE 4.2: **Estimated costs of and outputs from 64,000 hectares of eucalyptus plantations to feed a paper and paperboard mill with annual capacity of 30,000 tonnes**

Costs : Rs. 000
Outputs : 000 tonnes

Year	*Area planted : (hectares)*	*Plantation costs**	*O+M costs***	*Total annual costs (3+4)*	*Annual outputs*	*Present Value of Annual costs Discounted at 14 per cent*	*Present Value of Annual outputs Discounted at 14 per cent*
(1)	*(2)*	*(3)*	*(4)*	*(5)*	*(6)*	*(7)*	*(8)*
1	800	980.0	—	980.0	—	980.0	—
2	800	980.0	40.0	1,020.0	—	894.5	—
3	800	980.0	56.0	1,036.0	—	796.7	—
4	800	980.0	72.0	1,052.0	—	710.1	—
5	800	980.0	88.0	1,068.0	—	632.3	—
6	800	980.0	104.0	1,084.0	—	562.6	—
7	800	980.0	120.0	1,100.0	—	501.6	—
8	800	980.0	136.0	1,116.0	56.0	445.3	22.4
9	—	—	152.0	152.0	56.0	53.4	19.6
10	—	—	128.0	128.0	56.0	39.4	17.2
11	—	—	128.0	128.0	56.0	34.5	15.1
12	—	—	128 0	128.0	56.0	30.3	13.3
13	—	—	128.0	128.0	56 0	26.6	11.6

TABLE 4.2 (*Contd.*)

(1)	(2)	(3)	(4)	(5)	(6)	(7)	(8)
14	—	—	128.0	128.0	56.0	23.3	10.2
15	—	—	128.0	128.0	56.0	20.4	8.9
16	—	—	128.0	128.0	56.0	17.9	7.8
17	—	—	128.0	128.0	56.0	15.7	6.9
18	—	—	128.0	128.0	56.0	13.8	6.0
19	—	—	128.0	128.0	56.0	12.1	5.3
20	—	—	128.0	128.0	56.0	10.6	4.6
21	—	—	128.0	128.0	56.0	9.3	4.1
22	—	—	128.0	128.0	56.0	8.2	3.6
23	—	—	128.0	128.0	56.0	7.2	3.1
24	—	—	128.0	128.0	56.0	6.3	2.8
25	—	—	128.0	128.0	56.0	5.5	2.4
26	—	—	128.0	128.0	56.0	4.8	2.1
27	—	—	128.0	128.0	56.0	4.2	1.9
28	—	—	128.0	128.0	56.0	3.7	1.6
29	—	—	128.0	128.0	56.0	3.3	1.4
30	—	—	128.0	128.0	56.0	2.9	1.2
31	—	—	128.0	128.0	56.0	2.5	1.1
32	—	—	112.0	112.0	56.0	1.9	0.9
33	—	—	96.0	96.0	56.0	1.4	0.8
34	—	—	80.0	80.0	56.0	1.0	0.7
35	—	—	64.0	64.0	56.0	0.8	0.6
36	—	—	48.0	48.0	56.0	0.5	0.6
37	—	—	32.0	32.0	56.0	0.3	0.5

38	—	—	16.0	16.0	56.0	0.1	0.4
39	—	—	—	—	56.0	—	0.4
Total						5,885.0	179.1
Cost/price per tonne=Rs. 32.85							

*Estimated @ Rs. 1,22,500 per 100 hectares as in the case discussed in Table 4.1.

**Basis used in the preceding case (Table 4.1) not understood. Assumed @ Rs. 5,000 per 100 hectares in the 2nd year of planting and Rs. 2,000 per 100 hectares per year subsequently.

the yields per hectare cannot be uniform due to (i) differences in survival percentage at the harvesting age; (ii) differences in soil and climatic conditions; (iii) differences in the incidence of illicit cuttings, grazing, etc.; and (iv) differences in managemental quality/intensity. If the price per unit of output is determined in advance, as it should be, on the basis of *estimated* costs; the forestry system may not receive the expected returns. There may even be losses. If, on the other hand, price is determined on the basis of *observed* costs, the industry would have to bear the entire burden of losses on account of low yields resulting from inadequate protection, or insufficient management, or even inappropriate selection of the variety of a tree species.[81]

3) Cost-based pricing necessitates the choice of a discount rate. This has always been difficult, but additional complications can arise due to the prevalent administrative practices. For example in one case around the mid 70s, it was first agreed by all concerned that a reasonable rate of return on the investments made by a FD should be the basis for fixing the stumpage price of wood from a plantation crop. The state government constituted a committee to determine the reasonable rate of return. Representatives of the Forest and Industries Departments thought that a 10 per cent IRR should be deemed adequate, but the representative of the Finance Department insisted that it should be minimum of 15 per cent [Misra, 1980b, p. 5].

There was an acrimonious debate between the administrators mainly because one group perceived that its prime responsibility was to encourage the growth of forest-based industries while, the other thought of enhancing the flow of financial resources to the state exchequer. One of the administrators also wanted to ensure that any decision should not leave behind even an impression that there had been an effort to justify the suggestion made by another member [Misra, 1980b, p. 7]. There were prolonged discussions but the committee could not choose/recommend an IRR acceptable to all its members, and the matter had to be referred to an organisation of the GOI.

[81]It is reported that over 50 varieties of eucalyptus are available in the country. The forester must choose the most appropriate for a given site.

This is not to argue that appropriate care should not be taken in the choice of a discount (or compound) rate as that would have a significant bearing on the stumpage price. For the case discussed in Sub-section 4.2.9.1, for instance, stumpage prices at 10 and 12 per cent IRRs would have been around Rs. 52 and Rs. 62 per tonne, respectively, in place of Rs. 74 per tonne with the IRR at 14 per cent. The point is that different interest groups would have different interpretations of the words reasonable/adequate/sufficient, etc. At the same time, a uniform guideline for the choice of a discount rate can be made. The issue can be settled by adopting a general principle that the rate would be equal to the one charged by the National Bank for Agricultural and Rural Development (NABARD).

4) Yet another problem in using cost of production as the basis of pricing woody materials would arise from the practice of raising man-made plantations by clearfelling mixed natural forests. Situations can be visualised where returns from clear-felling mixed natural forests could meet a major portion of the initial costs of raising man-made plantations, and the practice of accounting the former as returns with the latter could mean extremely low cost per unit of output from man-made plantations.[82]

5) As discussed, the principle of cost-based pricing cannot be applicable to the outputs from natural forests. The production costs of this category of materials were neither known nor would they be easy to compute for various reasons including problems in determining the land rent. Some forest officers and administrators with whom we discussed the issue were also not in favour of using this principle because they stated that "our" investments in the natural forests have been negligible.

Some people on the contrary, have used the term *replacement costs* to argue that natural forests have degraded to a point of no return except through man-made plantations and, therefore, the produce from the former must also be priced on the basis of production costs from the latter. This, however, cannot be logical because the industry has not been held responsible for the poor state of India's forests. As a matter of fact, the

[82]This system of accounting was, of course, not logical. For details *see* : Gupta, 1979b.

silvicultural management of forest areas leased to the industry has been the responsibility of the FDs and they always had the authority to periodically revise or modify silvicultural rules and practices, and to ensure their due observance.[83]

It must also be recalled that the growth of the industry and the pattern of using cellulosic materials have resulted from policy interventions to attract manufacturing units to utilise the unutilised or underutilised natural forest produce, to widen the tax base, to accelerate infrastructural development, to generate employment opportunities in relatively backward/remote areas, etc. Instances of such policies can be found even now.[84]

The industry, however, looked at this from a different angle. They recognised that the forestry sector must recover the costs incurred to administer, protect, and improve natural forest areas through the sale of outputs, but argued that besides pulpwood and bamboo, natural forests also generated various other timbers which had higher commercial values. The FDs/FDCs have been receiving large revenues from timber, which have risen many-fold due to substantial rise in prices inr the recent past.

Moreover, the forests were being managed with the same zeal even before the industry started using bamboos for paper manufacture, and the pulpwoods are a by-product of the timbe production function of the forestry system. This implied that a rational method of apportioning the cost of managing natural forests amongst several products must be found. That would amount to unscrambling a scrambled egg and no one has found a way to do that. In general, this was a problem of pricing joint products which had rarely been solved to the satisfaction of all concerned.

[83]Only one agreement made in the 1930s between a government and a paper making company could be construed as an exception. It provided that silvicultural rules for the forest land under reference could be modified by the government/FD in consultation with the company.

[84]Some senior forest officers reasoned that if forest produce was to be supplied at relatively lower rates to attract investments in processing units, the state department of industries or industrial development corporations must compensate the forest departments. In one state of the Indian Union this was agreed in principle but the principle has never been practised.

In essence, if the principle of cost-based pricing is followed, prices of cellulosic materials from natural forests would have to be different compared with those from man-made forests. Administrative, managerial, and economic implications of the situation must be carefully assessed in advance.

6) The industry would also like to argue that while the cost of production should be a basis for pricing the outputs from man-made plantations, the prices should *not be equal* to the costs (of production). The reasoning is that even monoculture forests, if appropriately raised and managed, generate a number of intangible and indirect benefits which accrue to society : improvement in the quality of biophysical environment, enhanced protection for wildlife, enhanced water and soil conservation, etc. A portion of the total cost of production must be attributed to such societal benefits.

A major constraint with this reasoning is that very little attempts have been made at quantifying the indirect and intangible benefits, and for placing price tags on them. The industry's reasoning may, however, be countered as it also generates negative externalities borne by the society.

Our discussions during the course of this study, however, indicated that in spite of these problems the industry was willing to pay the production costs of the materials emanating from man-made tree plantations but would like that (i) only fast growing species should be raised at closer spacement on suitable sites; and (ii) the norms in respect of costs, yields and/or survival percentage of the harvesting age be predetermined. Some industrialists even thought that 90 per cent survival at the time of harvesting must be guaranteed.[85]

The industry has also argued that the costs per unit of output from plantations raised by the FDCs/FDs could be abnormally high because of the inefficiencies of the public systems. They would, therefore, like that land be leased to them to enable them to produce the required cellulosic materials. This is a basic policy issue.

4.2.9.3 Some Advantages in Using Cost of Production as a Basis for Pricing

The use of this principle would also have some long-term

[85]This can be counter productive if disproportionately large expenditure is incurred on protection and fencing.

advantages not only for the forestry system and the industry, but also for the economy in general.

1) Those responsible for formulation and execution of forestry development projects would have to be more conscious of their roles than has been the case. Some basic economic principles and norms would have to be observed. Concurrent and ex-post evaluation of the projects would be necessary. Productivity per unit of financial resources available for investment and of the natural resource systems can be expected to substantially improve.

2) The industry can be assured of regular supply of materials which can be handled, transported, stored, and chipped at relatively lower costs per unit simply due to uniformity in size; and can be better suited for pulping dueto the desired age. Considerable savings in financial and environmental costs of paper production could emerge.

3) Other factors remaining the same, the quality of final outputs can be expected to improve at the current costs or at marginally higher costs per unit.[86] The producers should be able to recover the incremental costs, and the consumers could also derive considerably higher satisfaction per unit of the expenditure on paper and paperboards.

4) Alternatively, substantial increase in cellulosic raw material prices resulting in considerable increase in pulp and paper production costs might be a step towards demand management. This may be desirable even though a conspicuous policy for that purpose can possibly not be adopted in the foreseeable future.[87]

4.3 Towards a Synthesis

The preceding discussions were sufficient to conclude that none of the suggested bases for pricing cellulosic raw materials of forest origin can adequately satisfy all the concerned parties. Some of the suggested bases, such as, the industry's capacity

[86]This can be more true with the desired changes in production technology.

[87]The realisation that the nation need not aim at a fast growth in per capita consumption of paper and paperboards appears to be growing, but forces to the contrary have also been active.

to pay, a predetermined share in profits of the paper and paperboard producers, a portion of prices of final products were neither logical nor practicable. At the same time, a combination of bases such as, past and current royalty rates, general price index, market price of comparable goods could be used for fixing the administered prices of the produce from natural forests in the short run, i.e., the period necessary to plan and raise industrial plantations up to maturity. The most logical and viable alternative in the long run, in spite of the problems discussed earlier, appeared to be to grow the materials for specific manufacturing units and to price them according to production costs.

This does not mean that the total demand for the materials must be met from man-made plantations. It has been reasoned that the outputs per unit of productive forest land can and should be enhanced and the effort did not necessarily have to be capital intensive (Chapter 3). This was also necessary to relieve the current tensions and to enable an objective assessment of the issues.

For pricing the outputs from natural forests, the forestry system can not insist on cost of production as a basis for reasons like (i) the difficulties in computing such costs, and (ii) the prospects of enhancing the outputs at relatively lower costs per unit. Variables such as past and current (not necessarily proposed) royalty rates, changes in prices of other inputs, changes in infrastructural facilities and productivity of forest areas under reference, changes in ex-factory prices of different grades of paper and paperboards, changes in the general price index, could be considered for the purpose.

Exercises for assigning weights to these variables can be done by the concerned parties. Alternatively, some independent organisations may be requested to undertake the work. A further discussion on the subject has not been possible for us as the perceptions of the forestry system and the industry were widely different, arguments between the parties had assumed the status of religious debates, judicial interventions had been sought in some cases, there had even been conscious and unconscious efforts at evading rather than facing the real issues, and the parties were not willing to share the relevant quantitative and qualitative information.

A sustainable long term solution would be to raise industrial plantations and to price the outputs on the basis of (but not necessarily equal to) production costs. A few essential points must be noted here.

1) The long run need not, in fact should not, be too long. The forestry system may like to continue to retain its status as a major supplier of cellulosic materials, and can prepare and implement the needed plantation projects within a few years.

2) Considerable portions of the woody materials for use by the pulp and paper producers could continue to emanate from natural forests, but fixing their prices need not continue to pose problems. Accumulated experiences of pricing the materials from man-made crops should be usable with adjustments made for differences in size and age of the materials where relevant, for differences in costs of harvesting and transportation, etc.

3) The plantations should be raised in consultation with, and for each of the pulp and paper manufacturing companies in the public and private sectors. This should be less difficult at the present juncture as a good portion of the installed capacity is unutilised and new units may not be needed for some time to come.

4) Prices of outputs from man-made plantations need not necessarily equal their production costs. Adjustments may have to be made to achieve certain wider objectives. For instance, vast scope appeared to exist for reducing the consumption of chemicals, coal, power, water, etc. per unit of the final outputs (Section 4.2.4). Considerable R & D efforts at the industry and institutional levels would be required to facilitate actions in the desired direction. Relative shares of the different entities in raising the required resources has been a ticklish issue. It has however, been argued that the manufacturing units which have long since reached the stage of zero depreciated value of the capital at hand should contribute relatively more to the R & D efforts. It remains to be examined whether this can be done through differential prices of cellulosic raw materials.

Another suggestion has been that the manufacturing units which consume relatively larger quantities of chemicals, power, water, etc., per unit of final outputs should be charged higher prices for cellulosic materials. This necessitates fixing the norms. Methods for appropriate adjustments in raw material prices to provide an incentive to the efficient producers may be examined along with other issues.

More importantly, adjustments may be necessary if the production costs are adjudged to be unreasonably high due to structural constraints. In such cases, organisations such as the industrial development corporations might find it desirable to subsidise the FDs/FDCs undertaking the plantation projects.

On the contrary, production costs per unit of output from some industrial plantations could be lower than what the industry can conveniently pay. This could happen if (i) prime forest land was used for raising the plantations, and (ii) the practice of accounting the costs of and returns from clearfelling natural forests as part of the costs and returns *with* the man-made plantations was continued.

5) Success and acceptability of the industrial plantations can be substantially enhanced by—

- —Selecting the sites in consultation with the industrial units;
- —Involving the users of the outputs while selecting the tree species;
- —Estimating the annual plantation and O and M costs as accurately as possible and seeking the industry's reaction to the reasoning behind the estimates;
- —Specifying the discount/compound rate usable for translating the cost streams to a common time period, and documenting the logic for making the choice;[88]
- —Stating the number of rotations and the duration of each rotation;
- —Indicating the estimated outputs per rotation and assuming/accepting the responsibility for any shortfalls;[89]
- —Adjusting the plantation and harvesting patterns to meet the concerned industrial units' annual demand for woody materials;
- —Harvesting according to the schedule so that computations of time value of money are not adversely affected; and

[88]The task is not easy. It has been suggested that the interest rate charged by NABARD may be accepted in principle. This would also ensure uniformity amongst states.

[89]Industry could be willing to accept adjustments on genuine grounds.

—Not insisting on accounting for such cost items which cannot be determined with the help of available logic and may not make a considerable difference in the production costs per unit.

The last of these points may be explained in the context of the problems in determining the rental value of forest land (Section 4.1.1). Some studies in the past also assumed the land rent at zero but that is not logical. On the contrary, a conscious decision to ignore the land rent may be justifiable as the users of the output would not receive direct and proportional benefits from the intangible goods generated by the plantations.

Similarly, the forestry system may not be keen to earn net surpluses or profits at least to start with. It would be difficult to find an agreement on 'reasonable' level of profits from or net return on the investments. It may even be argued that the system must honour the responsibility to supply the raw materials at a 'reasonable' rate to the industry.

The discussion can be concluded with the observation that the principle of cost-based pricing can have substantial merits in the long run, and the problems in practising it can be resolved if the forestry system and the industry take them as challenges and opportunities. The task can be further facilitated if the parties approach the issues in a business like manner, and objectively assess the wider social implications of their actions and their advice to policy makers.

The industry may also choose to further develop the alternative of entering into agreements with individual land-owners to supply cellulosic materials. A large number of farmers practically all over the country have been raising tree crops.[90] Institutional finance can be availed to facilitate the activity.

Financial feasibility of the activity and its acceptability by the farmers can be further enhanced through supply of effluent water for limited, say two to three per year, irrigations for the tree crops. The pulp and paper producers may also not continue to insist on eucalyptus, and may even encourage the farmers to raise multi-purpose tree species.

[90]Bamboo cultivation by individual farmers is also known and may be extended.

This is not to say that the industry's plea for lease of land to enable them to raise the raw materials for themselves should not be adequately assessed at the highest level. This is a policy issue. The policy decision may, however, be based on considerations that this is a variant of cost of production as the principle for pricing the woody materials. Its choice may also be necessary if the forestry system is unable to supply adequate quantities of the materials at predetermined prices for reasonably long periods of time. The alternative may also help in introducing healthy competition for objective specific and speedier rehabilitation of waste but productive land. Methods to synthesise the interests of the industry, the local people, and the economy can be found.

CHAPTER 5

Summary and Conclusions

Issues pertaining to demand-supply management and pricing policy of forest-based cellulosic raw materials for the writing and printing papers and paperboard industry in India have been of considerable importance, especially with the rise in annual outputs of paper and paperboards from just about 0.1 million tonnes in 1950 to a little over 1.5 million tonnes in 1986, rising competing demands on forest produce in general and bamboo and hardwoods in particular, dwindling natural forest wealth resulting in strong perceptions of absolute scarcity of woody materials in the near future, etc.

This was in complete contrast to the situation till a few years back when pricing the forest produce was not an important issue because the supply was considered to be almost inexhaustible. Apart from the age-old practices of open auctions or inviting sealed tenders for a large portion of the outputs including highly valuable timbers, and high level negotiations for fixing the royalty for materials to be supplied to the forest-based industries, there have been very little efforts at systematic marketing of wood and non-wood forest products.

For the supply of bamboo and mixed woods, the state governments and paper and paperboard manufacturing companies entered into formal and informal agreements. The agreements specified the royalty rates per tonne of bamboo and woods, the extent of periodic revision in royalty rates, procedures for weighment of the materials, duration of supply of agreed quantities of the materials, etc.

By definition, royalty can be fixed and charged for use of government property for harnessing a naturally occurring produce. In that sense, royalty was not the price of the forest produce but the two terms have been used interchangeably. Moreover, royalties were fixed on ad-hoc basis but the process was made out to be of great consequence so that considerable time was spent on negotiations whenever royalty rates were to be fixed and/or revised.

In spite of long drawn negotiations, the royalty rates of bamboo and mixed woods till the recent past have been rather low. The state governments have also been offering various financial concessions and facilities to the forest-based industries. The reasoning was that productive employment would keep the local people away from destructive practices vis-a-vis the forests, and the additional revenues accruing to the state by way of broadening the tax base could be ploughed back to enhance the productivity of forest areas. But these have rarely been valid.

Recent thinking on royalty rates of forest produce was substantially different from the past. In the initial stages of growth of the paper and paperboard industry only a few agreements provided for periodic upward revisions in royalty rates, and the revisions, where provided, stipulated very small increases. On the whole royalty rates of bamboo between 1940s and 1970s were revised at intervals of 5-10 years, and the rates thus revised were 10-20 per cent higher than the preceding levels. Contrary to this, the royalty rates prevailing during the early 1980s were 300 to 400 per cent higher than the rates prevalent during the middle of 1970s.

This, however, did not necessarily mean that the royalty rates in force have been logically determined. The task was not easy, and had been made more difficult by a confluence of variables including perceptions of acute scarcities of bamboo and mixed woods, rising prices of the traded portion of forest outputs in general and of primary timber in particular, an unprecedented desire on the part of the forestry system and the state governments to enhance revenues per unit of the forest outputs, fast growth of forest-based industries especially the pulp and paper industry, and inefficiencies in the production process, etc.

One consequence of the situation has been an environment of mistrust, confusion, and tensions. Another outcome has been that the sellers and buyers of forest-based cellulosic materials were indirectly testing their bargaining powers. These seemed to have obstructed objective assessments of the merits and demerits of alternative approaches to a pricing policy aimed at (i) resolving the felt and preceived conflicts amongst various categories of users of bamboo and mixed woods, (ii) managing the demand for forest products as well as manufactured goods without necessarily resorting to direct rationing, (iii) enhancing the flow of scarce products towards most efficient uses, (iv) encouraging self-sustained development of forests and other unutilised or underutilised land resources, (v) eliminating wide fluctuations in the supply and price of finished goods based on woody raw materials, etc.

The study was, therefore, undertaken with the objectives of:

—Assessing the potential demand for and supply of forest-based cellulosic raw materials for the paper and paper-boards industry in India.

—Assessing the logic and feasibility of the various bases suggested for pricing the cellulosic raw materials of forest origin.

—Attempting a synthesis amongst approaches for a long-term pricing policy of forest outputs used by the pulp and paper industry.

The study has not covered the newsprint industry. It was recognised that newsprint production competed with writing and printing papers, paperboards, and allied products for cellulosic raw materials and other resources. On the contrary, demand for newsprint has been only indirectly related to the demand for paper and paperboards; and almost the total newsprint consumption in India was accounted for by sections of the urban population which were relatively better off in terms of per capita income, educational level, etc. Price elasticity of demand for newspapers has also been low.

Based on an assessment of the past and current trends, it has been estimated that the demand for paper and paperboards in India may rise by 3.5 per cent per annum by A.D. 2000. This meant that, without any imports, the markets in the country should be in a state of near equilibrium if the production at home could gradually rise to 2.5 million tonnes by the turn of the century. However, in the context of the current installed capacity at nearly 3.0 million tonnes, production capacity in the pipe line, recent policies to encourage consumption, possibility of some exports, etc. it was considered desirable to work with a growth rate of 4.0 per cent per annum. This entailed a production of 2.7 million tonnes by the turn of the century.

It has been discussed that of the total estimated production of paper and paperboards in India by the year 2000, a maximum of 1.7 million tonnes would be based on conventional raw materials of forest origin, i.e., bamboo and hardwoods. There was almost a unanimity of thought that the mix of bamboo and hardwoods in the total would, at worst, be 40 : 60. Inputs of bamboo and hardwoods with the chemical pulp process were taken at 2.5 and 2.7 tonnes, respectively, per tonne of the final output. Thus the industry's demand for bamboo and mixed woods by the year 2000 would grow to a maximum of 1.7 million tonnes and 2.75 million tonnes, respectively. The input levels would be significantly lower with other pulping processes such as mechanical or chemo-mechanical. Changes in these directions were anticipated but have not been considered in this study because the approach has been to work with the higher side of the raw material requirements.

Assessments of the extent of productive forest area in the country, prospects of harnessing only a portion of the estimated productive potential through conventional and non-conventional measures of resource management, availability of financial and trained manpower resources, and availability of wastelands for tree plantations indicated that the objectives of enhanced outputs of timber, mixed woods, and bamboo to satisfy the competing demands could be met. Between bamboo and hardwoods, the prospects of raising the outputs of the latter appeared to be better in spite of the recent emphasis on forest conservation.

The task of supplying adequate quantities of bamboo and mixed woods to the industry would be further facilitated if other forest based materials such as grasses and reeds could support eight to ten per cent of the total outputs of the pulp and paper industry. We have, however, been conservative in this regard as the administrative and managerial feasibilities of large scale use of such materials have not been studied.

This should facilitate an objective approach at a self sustaining long run policy and practices for pricing (not royalty) the forests-based industrial raw materials. Logic and feasibilities of a number of approaches for pricing the cellulosic materials and for periodic revisions of such prices have been assessed in the context of accepted principles and practices for pricing economic goods and services. These included :

—Some past and current royalty rate(s),
—Payment of rent for the forest land,
—General price index,
—Capacity of the paper and paperboard industry to pay,
—Some share in profits of the paper and paperboard producers,
—Some portion of the prices of final products,
—Market price of the goods under consideration or of comparable goods, and
—Cost of production.

The assessments indicated that none of the suggested bases for pricing cellulosic raw materials would adequately satisfy the concerned parties. Some of these like, the industry's capacity to pay, a predetermined share in profits of the pulp and paper producers, a portion of prices of final products appeared to be neither logical nor feasible under the current socio-political and economic dynamics.

At the same time, a mix of bases such as past and current royalty rates, general price index, market price of comparable goods could be logically used for fixing the administered prices of the natural forest produce in the short run. For the long run the most logical and viable alternative could be to raise the materials for particular manufacturing units and to price them on the basis of, but not necessarily equal to, the production costs.

This should not mean that the total demand for the materials must be met from man-made plantations. The study has reasoned that the outputs per unit of productive forest land can and should be enhanced and the efforts do not have to be capital intensive. This was also necessary to relieve the current tensions and to facilitate an objective assessment of the issues.

It has been reasoned that prices of outputs from natural

forests should not be strictly based on production costs due to difficulties in computing such costs, and prospects of enhancing the outputs at relatively lower costs per unit. Variables such as past and current royalty rates, changes in prices of other inputs, changes in infrastructural facilities, changes in productivity of forest areas under reference, changes in ex-factory prices of different grades of paper and paperboards, changes in the general price index could be considered for the purpose.

Exercises for assigning weights to these variables can be done by the concerned parties or an independent organisation can be requested to handle the task. Further discussion on the subject has not been possible as the perceptions of the forestry system and the industry were widely different, arguments between the parties had assumed the status of religious debates, judicial interventions had been sought in many cases. There have also been conscious and unconscious efforts at evading rather than facing the issues, and the parties have not been willing to share the relevant data.

For the success of long run alternative of raising industrial plantations and pricing the outputs on the basis of production costs, a few essential points have been made:

1) The long run period need not be too long as the forestry system can prepare and implement the needed plantation projects within a few years to retain its status as a major supplier of cellulosic materials.

2) Considerable portions of the woody materials for use by the paper and paperboard manufacturers could continue to emanate from natural forests but fixing their prices need not continue to pose problems. Prices of the materials from appropriately planned and managed man-made plantations should be used with adjustments: for differences in size and age of the materials, for differences in costs of harvesting and transportation, etc.

3) The plantations should be raised for all the pulp and paper manufacturing companies both in the public and private sectors.

4) Prices of outputs need not necessarily be equal to the production costs. Adjustments would have to be made to achieve wider socio-economic objectives. In some cases, production costs per unit of output may even be lower than what the

industry could conveniently pay.

5) Sites and tree species for the plantations could be selected in consultation with the concerned manufacturing units.

6) Estimates of annual plantation, and O and M costs must be made as accurately as possible and the industry's reaction to the reasoning behind the estimates must be sought.

7) Discount/compound rate for translating the cost streams to a common time period must be specified and the logic for the choice must be documented. It would be better to adopt a uniform rate. The interest rate charged by NABARD can be a logical choice.

8) The number of rotations and duration of each rotation must be clearly stated along with estimated outputs per rotation. The project authority must bear the responsibility for any shortfalls in outputs but the industry must also be willing to accept visibly genuine adjustments.

9) Plantation and harvesting patterns must be adjusted to meet the annual requirements of the concerned industrial units and crops must be harvested according to schedule so that the computations of time value of money are not adversely affected.

10) Cost items which cannot be logically determined and not make considerable difference on the production costs per unit of the outputs (woody materials) need not be accounted. Rent of forest land and profits to the forestry system could be examples in this category.

Some specific advantages of the principal of cost-based pricing for the forestry system, the industry and the society have been discussed. It has been reasoned that the likely problems in practising the principle can be resolved if the concerned parties visualise the situation as one of challenges and opportunities, approach the issues in a business like manner, and objectively assess the wider social implications of their advice to the policy makers.

The industry may also choose to develop the alternative of entering into agreements with individual land-owners and operators for the supply of cellulosic materials. Farmers, practically all over the country have been raising tree crops. They can avail of institutional finance to further their activity. The task may be further facilitated through supply of effluent water for limited (say two to three per year) irrigation to the tree crops. This would enhance biomass production per unit of land and time. The pulp and paper producers may also not continue to insist on eucalyptus, and may even encourage the farmers to raise multi-purpose tree species.

The industry's argument for lease of land to enable them to raise woody materials for their own requirements also needs to be assessed at the highest policy making level. It was a variant of the cost of production principle. Its choice may also be necessary if the forestry system was not prepared to supply adequate quantities of the materials at the predetermined prices for reasonably long periods of time. The alternative may also help in introducing healthy competition for objective specific rehabilitation of the waste but productive land. Methods to synthesise the interest of the industry, the local people, and the economy can be found.

ANNEXURE 1

Some Basic Concepts and Issues in Pricing Economic Goods and Services

Price is one of the most frequently used terms by economists, policy makers, administrators, professional managers, business people, workers, and housewives alike. In spite of this, or perhaps due to it, the term does not have a distinctive or unique meaning. Discussions on the principles and policies of pricing economic goods and services have often been complicated and inconclusive. This has also been one of the major reasons for widely different recommendations by different individuals and/or groups and/or official committees on the basis or bases for pricing specific goods.

This annexure discusses a few basic concepts relevant for assessing the relative merits and demerits of alternative approaches to a pricing policy. The discussion would be of interest to those who are not well versed with economic and management concepts. The discussion would also help in a better appreciation of the problems in pricing forest-based cellulosic raw materials to the satisfaction of all concerned: policy makers, policy advisors, administrators, forest resource managers, industrialists, informed citizens and others.

Al.1 Demand and Supply

The concepts of demand and supply are pivots for any discussion on piece of a good/service. However, the term *demand* has been used interchangeably with *need* and *requirements*. There have also been studies to forecast the demand for paper and paperboards, and the demand for cellulosic raw materials of forest origin without any reference to their prices. This was an anathema to economists and professional managers. For them, demand meant *effective demand* and signified the quantity of a good or service that had been or was likely to be purchased/consumed in a given period of time and in a given situation. This implied that the desire of the prospective consumers to buy the commodity/service under consideration must be accompanied by their ability and willingness to pay for it. Requirements or wants of a good or service can assume the status of effective demand and vice versa with changes in its price.*

Similarily, the term supply of a good or a set of substitute goods indicated the specific quantity/quantities which the producers were willing or could be expected to sell at a given price and time. This, once again, could not necessarily synonymous with the total potential or actual production of the good under consideration. Supply of a commodity or service was known to change with changes in its price level.

AI.2 Distinguishing Features of Demand and Supply of Forest Products: Some Policy Implications

Demand for and supply of all the major forest products and a substantial portion of the non-wood (minor) forest products have a few distinctive features.

1) A very small portion of the total consumption of forest products such as timber, mixed woods, and bamboo can be classified as direct demand. A major segment of the total consumption of these goods emanated from indirect or *derived demand.* Most of the demand for timber, for instance, was derived from the need for the material for residential, institutional, and commercial complexes. This meant that the price of the final goods (semi-finished) for which the forest produce was to be used as a base material had to be an important consideration. Problems, however, arose as the long term prices of the processed goods and the consequent demand and supply functions could not be predicted with acceptable confidence levels.

*There are, of course, other factors affecting the levels of effective demand for and supply of a commodity or service. Those are not relevant to this discussion.

2) The derived demand characteristic of a product implied that the magnitude and the pattern of its demand could change significantly not only due to chances in its price and the prices of the finished goods for which it has been used as a base material but also, and perhaps more importantly, due to (i) changes in price, availability and acceptability of substitute base materials; and (ii) structural changes in the demand for finished products. Examples could be offered by changes in the nature and the mix of materials for residential and commercial buildings, changes in relative prices of substitute goods such as timber and steel, changes in consumption patterns of paper and paperboards, changes in relative prices and availability of substitute cellulosic raw materials, changes in pulping and paper making technologies, etc.

In general, however, the derived demand for a good would be more inelastic whenever—

—it was more essential,
—the supply of its substitutes was more inelastic,
—it accounted for a small proportion of total cost of the final product, and
—the demand for the final product was more inelastic [Gregory, 1972, p. 151].

A reflection on these variables in conjunction with the relevant facts about (i) the forestry system, and (ii) growth of the paper and paperboard industry in India could indicate that the demand for forest-based raw materials would be price inelastic in the short run. Due to inadequate reserves of the growing stock, the supply of forest products would also be price inelastic in the near future.*

These, in turn, could mean that the bargaining power of the forestry system was considerably stronger than that of the industry and that could be a part of the reasons for an often repeated statement that the forestry system in India has been, in the recent past, operating more like a monopolist producer

*It may be recognised that in situations of adequate reserves of growing stock, there can be one way flexibility in supply of forest products as the manager can have considerable choice in deciding the time for retaining or liquidating the capital in the form of trees. That, however, is not relevant to this discussion.

than a public-utility or a semi-commercial organisation which it has supposedly been. In essence, the position of the forestry system had changed from the one of a price taker to that of a price setter. Such a situation will have some impact on the long term pricing policy but the exact nature of such impacts would be a function of the perceptions of the concerned parties about their roles, advice rendered to the policy makers, etc.

3) Even the demand for certain types of forestry activities has in the past been and would continue to be derived from other natural resource management policies and activities. For instance, tree plantations on a substantial scale have been undertaken in association with projects aimed at watershed management in general and soil conservation and regulation of water movements in particular. Relative importance of these broader objectives should be one of the important considerations in pricing the tree based outputs. The task would, however, not be easy particularly because of situation specific variations in the relative importance of various objectives of such projects, and lack of adequate coordination amongst the developmental departments/executing organisations.

4) The distinction between demand and requirements drawn by economists and professional managers may not be valid for some of the forestry sector's outputs like fuelwood, fodder grasses, bamboo, etc. A good portion of these did not reach regular markets but were used by the local people as part of their *nistar* rights. Pricing policy in a demand-supply framework would necessitate that money values be assigned to the benefits accruing to the local people but that would not be easy or even desirable.

5) People can avail most of the intangible outputs of forest without expressing their demand and/or explicitly paying for them. Most forestry development projects, for example, would generally result in some improvement in the quality of the bio-physical environment. Those who express their demand for tangible outputs (timber, bamboo, mixed woods, etc.) have reasoned that the intangible benefits availed of by the society must be accounted for while fixing the prices of the tangible outputs. This, once again, was a difficult task.

AI.3 Market Price

In common parlance, price of a commodity or service was the payment made by the buyer to its seller. It was usually expressed in terms of money per unit (weight/numbers) of the commodity in question. When a good was not homogeneous, i.e., when its use value differed due to differences in intrinsic characteristics, market location, etc. the price per unit of the good would vary amongst its grades and/or locations.

When the quantity of a good offered by the sellers equalled the quantity taken by the buyers without any external restrictions or incentives, i.e., when the sellers and buyers were satisfied with the extent of the transactions in a given market during a given period of time, the traded price was labelled as the market price. If a particular price prevailed in the market for a reasonably long period of time, barring temporary fluctuations, it was called *normal price* and that usually indicated a harmony or near harmony between the divergent interests of the producers and consumers of the good.

It has been commonly believed that a producer wants to maximise net profits but that was not always true. A business firm may have other goals: (i) ensuring a pre-determined IRR on the investment, (ii) maximising sales revenues, acquiring market leadership, and (iii) establishing a distinct class or a clan of its own, not necessarily through maximising and accumulating net surpluses but, say, through maximising sales at lower price per unit of the product to reach a significantly larger number of people. Some of these objectives have been and could continue to be relevant for the forestry system in India.

Let it however, be maintained that profit maximisation was the basic objective of the business firms as it was the most common and the most pervasive of the goals to govern their behaviour. But contrary to the layman's thinking, maximum net profit could not be synonymous with highest price per unit of a firm's output(s). Principles of elementary economics would suffice to show that even a monopolist firm could not set very high price for its product and yet be able to earn maximum net profits.

Product prices were determined by a set of conditions under which the sellers and buyers operated. These set of conditions included the number of sellers and the number of buyers in a market, homogenity versus heterogeneity of the product, extent of constraints (or lack of them) on exit or entry of the producing firms, extent of knowledge on the part of the sellers and buyers about the nature of the product and its price, consumers' perceptions of intrinsic differences in the quality of close substitutes, etc. Based on the nature and variations in mix of such variables, economists have classified the product markets into different forms. Even with the profit maximising assumption, the pricing principles varied amongst the market types, i.e., the price which could be commensurate with the profit maximisation objective would vary with the market conditions.

This brief discussion would suffice for an understanding that what was commonly termed as the market price of a good or a service had different connotations not only between producers and consumers but also amongst their different groups. Questions such as :

—type of market : wholesale or retail, wide or restricted, etc.
—nature of the product : homogenous or heterogenous, relative ease in storing and transportation, etc.;
—economies or diseconomies of scale in production;
—relative bargaining power of the sellers and buyers; and
—extent of information about the price of different grades of the product;

must be answered in any professional discussion.

These points could also suffice to show that the market price(s) of a product can be a possible basis for fixing its administered price only if a *necessary* condition that neither the sellers nor the buyers were price setters or price takers was fulfilled. This has rarely been valid between the sellers and the buyers of forest-based pulping materials. Even at the cost of some oversimplification, it can be said that ever since its birth till about the middle of the 1970s, the pulp and paper industry in India was the price setter for these materials, but by late 1970s or early 1980s, the state governments/FDs had taken that position.

It must also be noted that there have not been many instances of pricing the forest-based cellulosic materials equal to the market prices of comparable or similar goods. However, it may be recalled that in one of the three states selected for a somewhat detailed study of the issues, market price of hardwoods was accepted as one of the relevant considerations for fixing the royalty of the material supplied to a paper making company around the middle of 1970s. Till that time, fuelwood and charcoal merchants were the only bulk buyers of the material and the average price realized by the forestry system was Rs. eight to nine per standard stack measuring 2 cm(r). Following negotiations between the parties, the royalty per stack payable by the paper making company was fixed at Rs. 16. The reasoning was that the market price depended on and varied widely amongst the forest coupes for several reasons including—

—location of the material vis-a-vis the major markets;
—solid contents of the fuelwood stacks;
—proportion of small and large sized billets in a fuelwood stack;
—state of the charcoal market; and
—time of the year (winter/summer) when the material was offered for sale.

Besides these reasons for modifying the average market price, the paper making company also recognised some additional advantages such as availability of large quantities of hardwoods in a concentrated area close to their mills, and relatively lower total landed cost compared to that of bamboo in spite of low royalty rates of the latter.

Thus, the market price was used merely as the starting point for the negotiations between the government and the company. The company's consent to pay more than the average price realised from the fuelwood and charcoal merchants could perhaps not be possible in the absence of an appreciation of mutual advantages to both the parties and, more importantly, without an environment of mutual trust.

AI.4 Administered Prices

Prices of commodities and services were not always determined by the market forces and/or were not necessarily related to their production costs. Prices of some goods or services have been and may be pre-determined for various reasons including imperfections in the market, a reasonably good estimate of total supply and demand for a product at the predetermined price. These were generally known as administered prices.

There were a number of bases for arriving at the administered prices. These varied not only amongst good but also for the same good at different points of time or in different situations. The important ones were—

—need based pricing;
—cost based pricing; and
—market forces based pricing.

AI.4.1 Need Based Pricing

Conceptually, need based pricing can be relevent only to centrally planned economies. In practice, however, this system has also been used in the free enterprise and mixed economies. This has been particularly relevant with the increasing acceptability of the welfare state tenet. Administered prices of social services: health care charges, school/college fees, local transportation fares, etc. were some of the examples.

Not much systematic thinking appeared to have been done to place this basis of pricing on an objective keel. Income gradations has been the most commonly used approach for pricing health care and educational facilities, but there have been problems in administering it. The organisations responsible for pricing the services and those providing income generating avenues had little regular linkages. The basis, therefore, necessitated income verification procedures which were often time consuming and yet did not yield reliable results.

Moreover, it was widely believed that this system has had adverse impacts on the quality of the services, and on the management of the equipment and other resources used in the process. It has also been observed that provision of services at the predetermined and subsidised prices diverted the consumers away from the relatively more efficient systems which, in turn, further enhanced the inefficiency of the subsidised systems as well as the need for subsidies.

Another reference point for need based pricing has been an estimated reasonable market price where (i) a public agency was responsible for a major portion of the total supply of the good/

service under consideration, (ii) its supply was assessed to be short of the requirements (not necessarily demand), and (iii) it was expected to serve a basic socio-economic need, and its cost of production was higher than the perceptions of a reasonable price. A good example could be prices of processed milk supplied by municipal dairies with or without regulating the quantity which a family unit could purchase. The logic has been that standardising the quality of a basic source of nutrition would have higher and long run social benefits compared to the financial losses due to subsidised prices.

Another example could be the bamboo supplied by the FDs to the local people at substantially lower or even negligible prices compared to the market prices. The arrangements have been justified on the reasoning that those who lived near or inside the forest areas were entitled to such quantities of the produce as would suffice to meet their household requirements and to support their economic activities through the cottage industries. It has also been argued that such steps induced the local people to protect the forest wealth.

A second basis for need based pricing could be provided by a prevailing norm or practice. One example was offered by food prices in the institutional cafateria. Another example, to repeat, was offered by prices of forest products supplied to or availed by the local people. Free or almost free supply of the forest produce to the people to meet their bonafide domestic needs has been practised. Head loads of fuelwood and specified quantities of bamboo taken by the local people under the *nistar* rights belonged in this category.*

Need based prices of certain goods and services could also be fixed through an assessment of the society's poorest section's capacity to pay. In such cases, however, quantity of the good(s) supplied at the administered prices has generally been restricted. Prices of levy sugar, certain types of foodgrains, and the *janata* (common) cloth have been based on such considerations. At the same time, systematic exercises of the people's capacity to

*Concerns have long since been voiced to progressively abolish *Nistar* but the socio-political dynamics favour its continuation. To make a case for or against such supplies of forest produce, or to assess their impact on forest resource management was not an objective of this study.

pay for such goods and services have been conspicuous by their absence.

The discussion in this sub-section would suffice to conclude that need based pricing can serve important social and economic objectives but necessitates objective assessments of the supply demand situations of the goods under consideration. This, in turn, could be necessary to ensure that the pricing policy itself did not aggravate a situation of perceived or real scarcities. Moreover, administration of need based pricing was not an easy task. The mechanisms in use have resulted in malpractices such as false declarations of incomes, unconscious and conscious deterioration in the quality of goods, diversion of the product(s) to markets for which they were not assigned, etc. Problems also arose due to difficulties in estimating the demand for a product priced on the basis of its need. Subjective assessments of requirements had to dominate the decision making processes such that analyses of the accumulated experiences could not be useful for long term policy making.

AI.4.2 Cost Based Pricing

This system of pricing has been or could be relevant in situations of contract selling of final outputs to governments, selling the outputs of a project which must break-even for whatever reasons, etc. A number of alternatives such as average cost pricing and marginal cost pricing including a mark-up for profits have been used within the principle of cost based pricing. These would generally lead to different results. An important issue in operationalising this principle would, therefore, relate to the choice amongst the production cost concepts. This has been and can be a difficult issue even though the behaviour of costs was usually better known than the behaviour of market price of a good.

Average cost pricing has been a time honoured technique. It necessitated identification and measurement of all costs* including a normal or fair margin, and division of the total cost by the number of units of the good produced. For this reason, it was also known as full cost pricing or average cost plus mark-up

*Indirect and intangible costs were usually not accounted for by the producing firms.

pricing. Conceptually, it can be a logical basis for running a business enterprise or a public utility service. In practice, however, there could be several limitations.

1) It implied that resources of all the firms in an industry were optimally utilised and there was no significant difference between a standard cost of production for the industry and the costs incurred by each firm. This was rarely valid.

2) Assurance of the recovery of full costs and normal profits could leave the producers with no incentive (or even a desire) to enhance the efficiency of the production process. Even the routine R and D activities to test and adapt new technologies could suffer. The problem would be further complicated in a situations of sub-optimal equilibrium where the IRRs per unit of the investments by industrialists have been or could be high, but returns from the scarce natural resources such as land have been extremely low.

3) It would generally be difficult to adequately and objectively estimate the cost of all the inputs used by a production organisation. In forestry operations, for instance, determination of economic rent has not only been difficult but may also not be logical (Chapter 4). Similarly, it has been difficult to precisely define normal or fair profit margins. In economic parlance, production costs included normal profits, and profits above the normal were classified as rent or surplus. It could, therefore, be reasoned that the annual revenues accruing to the forestry system included profits as they have been substantially higher than the annual costs, and the land earmarked as forests cannot be used for any other purpose. At the same time, the need for enhancing the administered prices of forest outputs has not been denied. This situation could raise a number of questions such as :

—Should normal profits be conceived as some percentage of investment ?

or

—Should normal profits be a pre-determined portion of the salaries paid to the forest officers or of total annual costs?

or

—Should normal profits be a fixed amount per unit of the land ?

These questions would not be easy to answer. Moreover,

the choice of any one of the stated variables as the basis for normal profits would have different policy implications.

It can even be reasoned that the forestry system in India should not necessarily be concerned about the extent of net profits. Quite a few enterprises in the public sector have long since operated and continued to operate in losses, and the surpluses generated by the forestry system were not invested to enhance the productivity of forest land even though its societal benefits were substantially higher than many other economic activities. The reasoning may appear to be counter productive, but the essence is that profit levels may not be a major consideration in pricing the forest outputs unless the policy was to place the forestry system on a self-sustaining commercial footing.

4) The concept of cost based pricing assumed that a firm produced a single homogeneous good. This was rarely valid. Forests generated various types of wood and non-wood products, and the quality and mix of these outputs had to vary with site quality, climatic conditions, management objectives, etc. Moreover, a number of intangible societal benefits: soil and water conservation, ameliorating effects on the bio-physical environment, etc. emanated from forestry activities.*

Very little attempts, if any, have been made to quantify the intangibles and to place situation specific price tags on them. At the same time, the consumers of tangible outputs can argue that the net societal benefits must be accounted while determining the prices of outputs supplied to them. This was almost impossible because costs cannot be satisfactorily allocated amongst a host of tangible and intangible, and direct and indirect outputs.

The problem can also not be resolved by using normative concepts. A standard cost of production, for example, cannot be objectively assessed due to significant variations in maturity periods of tree crops caused by variations in management practices, variations in soil and climatic conditions, significant variations in the quality and extent of tree cover on the forest land, institutional/legal/social/organisational constraints on free play of the forces affecting production efficiency, etc.

*There may also be negative externalities from man-made forests or monoculture.

Marginal cost pricing was another time tested concept in the economists' dictionary. Except in the short run, business firms generally tried to optimise net revenues over and above the normal profits. A firm's net profits would be optimum at that stage of production where the marginal cost (MC) was equal to marginal revenues (MR). The condition of MC=MR must be met in that segment of the production zone where the costs per unit of output are increasing, i.e., the law of diminishing marginal returns was in operation.

On the contrary, some organisations responsible for providing public utilities such as water and electricity were known to consciously run in losses. But, that cannot be in the best long run interest of an economy. Therefore, the long run objective of such organisations could be to break-even. This is synonymous with pricing the utilities to fully recover the production cost.

At the cost of some over simplification it may, therefore, be said that individual firms generally tried to optimise their net revenues and produced up to the point where MC=MR while, public utility firms could be using the principle of AC=AR. This may be explained with the help of an example. For the sake of simplicity, it may be assumed that MC and AC are linear functions.

Figure A.1 depicts the usual cost and revenue functions. A business firm operating to optimise net profits was expected to work on the principle of MC=MR (point K on the diagram) and would have, thus, produce 26,000 units of the good under consideration. The average price and average cost at that level of output would be Rs. 93 and Rs. 55, respectively. Thus, the total net profits would be—

$$\begin{aligned} &\text{Rs. } (93-55)\times 26{,}000 \\ &=\text{Rs. } 38\times 26{,}000 \\ &=\text{Rs. } 988{,}000 \end{aligned}$$

On the other hand, a public utility firm desirous of recovering the full costs but not of earning net profits would operate up to the point where AC=AR (P on the diagram) and produce 48,000 units of the good.

Both these situations, where one type of firms operated on

MC=MR principle while the others were guided by the AC=AR consideration would not indicate most appropriate utilisation of the productive resources. In the case of the public utility firm,

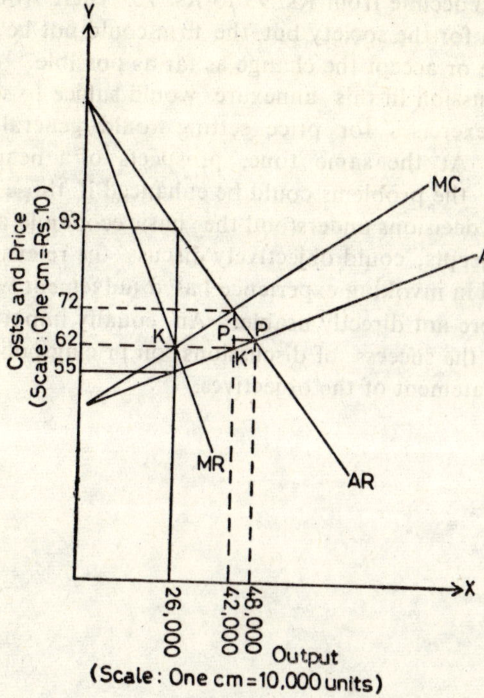

Fig. A.1.

for example, MC for a portion of the total output represented by the segment PP (=approximately 6,000 units of the good) was higher than the average revenued. This amounts to over utilisation of the productive resources at the firm's command. On the contrary, the business firm earned optimum net profits when the output was 26,000 units. At that level, however, the marginal cost was lower than the average revenue. This implied that the firm could enhance its total output from 26,000 to 42,000 units and yet earn a net surplus equal to :

Rs. (72—62) × 42,000
=Rs. 10 × 42,000
=Rs. 420,000

The net profits in that case would be lower than the optimal level by Rs. 5,68,000. At the same time, the total output would increase from 26,000 to 42,000 units, and the price per unit of output would decline from Rs. 93 to Rs. 72. There would be an obvious gain for the society but the firm could not be expected to appreciate or accept the change as far as possible.

The discussion in this annexure would suffice to show that conceptual exercises for price setting would generally be not practicable. At the same time, prospects of a near optimal solutions to the problems could be enhanced if those involved with pricing decisions understood the basic economic and management concepts, could objectively discuss the relevant issues, and believed in invoking experience based judgement when facts and logic were not directly usable. An equally important prerequisite for the success of discussions on pricing policy would be a clear statement of the objectives.

References

Agarwala, J. C. 1978. "Development of Paper Industry : New Strategy Vital", *The Economic Times*, September 2.

Agarwala, V.P. 1982. "Forest Raw Materials for Paper and Pulp Industry", in *Papers Presented at the Seminar on Raw Materials for the Paper Industry*, Indian Paper Makers Association (IPMA), Calcutta, pp. 42-50.

Ahuja, S.P. (Editor). 1980. *Paper Industry in India : Retrospect and Prospects*, Institute of Economic and Marketing Research, (IEMR), New Delhi.

Anonymous. 1960. "Marked Expansion in Paper", *Journal of Industry and Trade*, 10 (9) : 1420.

———. 1961. *Hundred Years of Indian Forestry*, Forest Research Institute (FRI), Dehradun.

———. 1965a "Bagasse for Paper Production in the South", *Eastern Economist*, 45 (16) : 740-742.

———. 1965b. "Paper Industry in India : A Study of Raw Materials Consumption", *Journal of Industry and Trade*, 15 (1) : 62-64.

———. 1966. *All India Bamboo Study Tour and Symposium Papers : 1963-64*, FRI, Dehradun.

———1967. *Preinvestment Survey of Forest Resources : Why, What, When, How?*, UNDP and GOI, New Delhi.

———. 1968. *Proceedings of the Symposium on Utilization of Hardwoods for Pulp and Paper*, FRI, Dehradun.

———. 1971. *Conference on Utilization of Hardwoods for Pulp and Paper*, FRI, Dehradun.

———. 1972. *Problems and Instruments Relating to the Allocation of Environment Costs*, Organisation for Economic Cooperation and Development, (OECD), Paris.

———, 1975a. "Paper Industry-I : Production Process and Technology", *The Financial Express*, August 29.

———. 1975b. "Paper Industry-II : Structure and Growth", *The Financial Express*, August 30.

———. 1975c. "Paper Industry-III : Price Control and Finance", *The Financial Express*, September 1.

———. 1975d. "Paper Industry-IV : Prospects and Policies", *The Financial Express*, September 2.

———. 1977. *Export and Import of Wood, Wood Products and Forest Resources*, FRI, Dehradun.

———. 1978a. "Economics of Expansion of Existing Units : Paper Industry", *Commerce*, 137 (3500) : 65-69.

———. 1978b. "Hindustan Paper Corporation : Special Feature", *Lok Udyog*, 12 (8) : 51-56.

———. 1978c. "Paper Industry : A Feature", *Commerce*, 137 (3500) : 65-71.

———. 1979. "Paper and Paper Board : An Industry Profile", *Industrial Research*, 5 (4) : 193-198.

———. 1980a. "Paper Industry : Position and Prospects-I" *SBI Monthly Review*, Number 1941, pp. 16-20.

———. 1980b. "Paper Industry : Position and Prospects-II" *SBI Monthly Review*, Number 1942, pp. 71-78.

———. 1980c. "Paper Industry : Position and Prospects-III" *SBI Monthly Review*, Number 1943, pp. 104-115.

———. 1981. "Paper and Paper Products : A Feature", *Commerce*, 143 (3660) : 1-23.

———. 1986. *Nistar Pustika Varsh 1985-86* (Hindi), Madhya Pradesh Government, Forest Department, Indore. Forest Division, Indore,

Arne, Sundelin. 1963. *Pulp and Paper Prospects in Western Europe*, FAO, Rome.

Arora, V.K. 1982. "Reclaiming Waste Materials for Pulping : Non-Wood Fibres", in *Papers Presented at the Seminar on Raw Materials for the Paper Industry*, IPMA, Calcutta, pp. 198-202.

Arya, D.D. 1968. "A Note on the Establishment of a Paper Mill in Rajasthan", in *Proceedings of the Symposium on Utilization of Hardwoods for Pulp and Paper*, FRI, Dehradun, pp. 42-44.

Aslam, S.M.A. 1968. "Rubber Wood as an Alternative for Commercial Pulp in the Manufacture of Paper", in *Proceedings of the Symposium on Utilization of Hardwoods for Pulp and Paper*, FRI, Dehradun.

Bachkheti, N.D. 1982. "Forest Resources: Keynote Address", in *Proceedings: Seminar on Raw Materials for Paper Industry*, IPMA, Calcutta, pp. 26-31.

Balasubramania, R. 1981. "Paper : Some Aspects of Taxation", *The Management Accountant*, 16 (7) : 313-314.

Banerjee, A.K. 1982. "Some Systems Constraints in Forest Management", in *Papers Presented at the Seminar on Raw Materials for the Paper Industry*, IPMA, Calcutta, pp. 72-74.

Banthia, K.M. 1976. "Cost Reduction in Paper Units : Concept and Scope", *The Economic Times*, December 22.

Basu, Bidyut Kumar. 1979. "Problem of Paper Trade", *Commerce*, 139 (3569) : 934-935.

Batra, V.P. 1981. "Growth Potential of Paper Industry", *The Financial Express*, May 18.

Bhandari, I.M. 1971. "Hardwoods for Paper : Are We Really Serious?", in *Conference on Utilization of Hardwoods for Pulp and Paper* FRI, Dehradun, pp. 64-68.

Bhatia, C.L. *et al*. 1982. "A computerized Approach for Pulp and Paper

Industries Development", in *Papers Presented at the Seminar on Raw Materials for the Paper Industry*, IPMA, Calcutta, pp. 75-83.

Bilgrami, I.A. 1979. "Capacity Utilization in the Paper Industry: Need for Incentives and Reliefs", *National Investment and Finance Weekly*, 37 (20) : 15661-63.

Biswas, Souren. 1981. *Presidential Address*, Annual General Meeting of the Indian Paper Mills Association, Calcutta.

———. 1982a. "Industrial Plantations : Keynote Address", in *Proceeding: Seminar on Raw Materials for Paper Industry*, IPMA, Calcutta, pp. 53-56.

———. 1982b. "Need for Industry Oriented Plantations", in *Papers Presented at the Seminar on Raw Materials for the Paper Industry*, IPMA, Calcutta, pp. 133-136.

Chaliha, B.P., Lodh, S.B. and Iyengar, M.S. 1966. "Development of Pulp & Paper Industry in India with Special Reference to Raw Materials", in *CSIR Working Group : First Get-together of Research & Industry*, CSIR, New Delhi.

Chaturvedi, B.N. 1982. "Forest Resources : Observations", in *Proceedings : Seminar on Raw Materials for Paper Industry*, IPMA, Calcutta. p. 50.

Chaudhry, Uma Dutta. 1977. "Industrial Breakdown of Capital Stock in India", *The Journal of Income and Wealth*, 1 (2).

CRB (Commerce Research Bureau). 1979. "Paper Industry : A Profile" *Commerce*, 139 (3569) : 923-929.

CSE (Centre for Science and Environment). 1982a. *The State of India's Environment 1982 : A Citizen's Report*, CSE, New Delhi.

———. 1982b. "Forests Mismanaged", *New Scientist*, November 25, pp. 490-495.

Chugh, K.L. 1982. "Rejuvenation of Bamboo Forests", in *Papers Presented at the Seminar on 'Raw Materials for the Paper Industry'* IPMA. Calcutta, pp. 63-71.

DCPPI (Development Council for Paper, Pulp and Allied Industries). 1983a. *Report of the Sub-Committee on Problems of Small Paper Mills with Special Reference of Recovery*, DCPPI, New Delhi, (Mimeo).

———. 1983b. *Investments for Paper Industry*, DCPPI, New Delhi. (Mimeo).

———. 1983c. *Energy Audit and Conservation in Pulp and Paper Industry in India*, DCPPI, New Delhi, (Mimeo).

———. 1983d. *Report of the Raw Materials Committee*, DCPPI, New Delhi.

———. 1983e. *Demand Forecast on Paper and Paperboard up to 2000 AD*, DCPPI, New Delhi.

———. No Date. *Formulation of Targets for the Fourth Plan and Perspective Planning for 15 Years: 1966-81*, DCPPI, New Delhi.

Deepak, K. 1978. "The plight of Indian Paper Industry : A Way Out", *National Investment and Finance Weekly*, 36(36) : 2482-85.

Deka, J.N. 1981. "Jute Production and Paper Industry", *Yojna*. 24(24) : 13-14.

Dhanuka, H.P. 1973. "Joint Sector Approach May Help", *Commerce*, 126(3237) : 1-3.

Dhanuka, S.S. 1982. "Problems Relating to Availability of Basic Raw Materials for Paper Mills in West Bengal", in *Papers Presented at Seminar on Raw Materials for the Paper Industry*, IPMA, Calcutta, pp. 109-112.

Dholakia, Bakul H. 1983. *Behaviour of Capital Output Ratios in Indian Economy*, Indian Institute of Management, Ahmedabad (Mimeo).

Duerr, William A. 1960. *Fundamentals of Forestry Economics*, McGraw Hill, New York.

Eddison, John C.A. 1955. *A Case Study in Industrial Development : The Growth of the Pulp and Paper Industry in India*, Centre for International Studies, Massachusetts Institute of Technology, Cambridge.

ESRF (Economic and Scientific Research Foundation). 1969. *Pulp and Paper Prospects for 1975 : A Study of the Basic Raw Material Requirements of the Indian Pulp and Paper Industry*, Natural Resources and Industrial Development Series, ESRF, New Delhi.

FAO (Food and Agriculture Organisation of the United Nations). 1960. *World Demand for Paper to 1975*, FAO, Rome.

———. 1962a. *Proceedings of the Conference of Pulp and Paper Development in Asia and the Far East : Volume I*, FAO, Bangkok.

———. 1962b. *Proceedings of the Conference of Pulp and Paper Development in Asia and the Far East : Volume II*, FAO, Bangkok.

———. 1963. *Pulp and Paper Prospects in Western Europe*, A study under the Direction of Arne Sundalin, FAO and BLV VERLAGSGESELLSCHAFT, Munich.

———. 1970a. *Pre-Investment Survey of Forest Resources India : Yield and Costs of Industrial Wood and Bamboo*, UNDP-FAO, Rome.

———. 1970b. *Pre-Investment Survey of Forest Resources India: Forest Resources Survey*, UNDP-FAO, Rome.

———. 1977a. *World Pulp and Paper : Demand, Supply and Trade-Volume I*, Selected Papers of an Expert Consultation Held in Tunis, 20-22 September, FAO, Rome.

———. 1977b. *World Pulp and Paper : Demand, Supply and Trade-Volume II*, Selected Papers of an Expert Consultation held in Tunis, 20-22 September, FAO, Rome.

———. 1980a. *Pulping and Paper Making Properties of Fast Growing Plantation Wood Species : Volume I*, Forest Industry Division, Forestry Department, FAO, Rome.

———. 1980b. *Pulping and Paper Making Properties of Fast Growing Plantation Wood Species : Volume II*, Forest Industry Division, Forestry Department, FAO, Rome.

———. 1981a. *Forest Resources of Tropical Asia*, Tropical Forest Resources Assessment Project, FAO, Rome.

———. 1981b. *Production Year Book 1980*, FAO, Rome.

———. 1985a. *Pulp and Paper Capacities Survey 1981-86*, FAO, Rome.

———. 1982b. *World Forest Products Demand and Supply : 1990 and 2000*, FAO, Rome.

———. 1983. *Year Book of Forest Products : 1970-81*, FAO, Rome.

FAO and ECAFE. 1962a. *Pulp and Paper Prospects in Asia and the Far East ; Volume II*, FAO, Rome.

———. 1982b. *Pulp and Paper Prospects in Asia and the Far East : Volume II*, FAO, Rome.

Francois, 1950. *Forest Policy, Law and Administration*, FAO, Rome.

Gadgil, Madhav *et al.* 1982. *Some Aspects of Forestry in India*, Centre for Monitoring Indian Economy, Bombay.

Ghosh, B. 1980. "Forest Raw Materials for Paper Industry", *Commerce*, 141 (3619) : 7-8.

GOG (Government of Gujarat). 1978. *Annual Administration Report of the Forest Department for the Year 1974-75*, Government Press, Rajkot.

———. 1979. *Annual Administration Report of the Forest Department for the Year 1975-76.* Government Press, Rajkot.

———. 1979a. *Annual Administration Report of the Forest Department for the Year 1976-77*, Government Press, Bhavnagar.

———. 1979b. *Annual Administration Report of the Forest Department for the Year 1977-78*, Government Press, Bhavnagar.

———. 1982. *Annual Administration Report of the Forest Department for the Year 1978-79*, Government Press, Vadodara.

———. 1982a. *Annual Administration Report of the Forest Department for the Year 1979-80*, Government Press, Vadodara.

———. 1982b. *Annual Administration Report of the Forest Department for the Year 1980-81*, Government Press, Vadodara.

Gohel, V.P. 1982a. "Industrial Plantations : Observations", in *Proceedings : Seminar on Raw Materials for Paper Industry*, IPMA, Calcutta, pp. 62-65.

———. 1982b. "Need for Plantations of Fast growing Species", in Papers Presented in Seminar on Raw Materials for the Paper Industry, IPMA, Calcutta, pp. 137-140.

GOI (Government of India). 1953. *First Five-Year Plan*, Planning Commission, GOI, New Delhi.

———. 1956. *Second Five-Year Plan*, Planning Commission, GOI, New Delhi.

———. 1959. *Tariff Commission Report on the Fair Ex-Works and Fair Selling Prices of Paper Boards*, GOI, Tariff Commission, Bombay.

———. 1962. *Third Five-Year Plan*, Planning Commission, GOI, New Delhi.

———. 1965. *Study on Forest Raw Materials for Pulp, Paper and Newsprint*, Committee on Natural Resources, Planning Commission, GOI, New Delhi.

———. 1970. *Fourth Five-Year Plan*, Planning Commission, GOI, New Delhi.

———. 1972. *Interim Report of the National Commission on Agriculture on*

Production Forestry : Man-Made Forests, Ministry of Agriculture, GOI, New Delhi.

———. 1973. *Interim Report of the National Commission on Agriculture on Social Forestry*, Ministry of Agriculture, GOI, New Delhi.

———. 1974. *National Commission on Agriculture : Interim Report on Desert Development*, GOI, New Delhi.

———. 1976a. *Fifth Five-Year Plan*, Planning Commission, GOI, New Delhi.

———. 1976b. *Report of the National Commission on Agriculture Part III : Demand and Supply*, Ministry of Agriculture and Irrigation, GOI, New Delhi.

———. 1976c. *Report of the National Commission on Agriculture Part IX: Forestry*, Ministry of Agriculture and Irrigation, GOI, New Delhi.

———. 1978a. *Medium Term Plan Forestry Sector : 1978-83 Report of the Working Group on Forests*, Department of Agriculture, Forest Division, GOI, New Delhi.

———. 1978c. *Final Report of the Sub-Committee on Pulp and Paper Industry*, GOI, New Delhi (Mimeo).

———. 1978c. Letter no. 1-1/78-FRY (FD) dated 13-10-1978, *Suggested Approach to the Pricing of the Raw Material to Forest Based Industries*, Ministry of Agriculture and Irrigation, Department of Agriculture, New Delhi.

———. 1980. *Scheme for Pasture, Fodder and Fuel Development on Private Lands*, Ministry of Rural Reconstruction, GOI, New Delhi.

———. 1981a. *Sixth Five-Year Plan*, Planning Commission, GOI, New Delhi.

———. 1981b. *Report on Development of Backward Hill Area*, National Committee on Development of Backward Areas, Planning Commission, GOI, New Delhi.

———. 1983a. *Report of Committee on Orientation of Forest Education in India*, Ministry of Home Affairs, GOI, New Delhi.

———. 1983b. *Economic Survey : 1982-83*, Controller of Publications, Delhi.

———. 1984a. *Report of the Working Group to Explore the Possibilities of Raising Industrial (Pulpwood) Plantations*, Ministry of Agriculture, GOI, New Delhi, (Mimeo).

———. 1984b. *Economic Survey : 1983-84*, Controller of Publications, Delhi.

———. 1985a. *Seventh Five Year Plan : 1985-90 Volume I*, Planning Commission, GOI, New Delhi.

———. 1985b. *Seventh Five Year Plan: 1985-90 Volume II*, Planning Commission, GOI, New Delhi.

GOK (Government of Karnataka). 1982. *Order No.* FFD-SI, FD/81, March 1, Bangalore.

Gray, John W. 1983. *Forest Revenue Systems in Developing Countries: Their Role in Income Generation and Forest Management Strategies*, FAO, Rome,

Green, Halgh P. and Palmer, E.R. 1985. "Technical Considerations Affecting the Minimum Size of Pulp and Paper Mill", *Indian Pulp and Paper*, 38(6); 15-21.

Gregory, G. Robinson. 1972. *Forest Resource Economics*, Ronald Press, New York.

Guha, S.R.D. and Singh, Man Mohan. 1960. "Chemical Pulps for Kraft Papers from a mixture of Conifers", *Indian Forester*, 86(9) : 563-65.

Guha, S.R.D. *et al.* 1961. Chemical Pulps for Wrapping Papers from a Mixture of Broadleaved Woods", *Indian Forester*, 87(3) : 200.

Guha, S.R.D. *et al.* 1968a. "Utilization of Eucalyptus Hybrid (Mysore gum) Grown in U.P. for the Production of Wrapping Paper Writing Paper and Rayon Grade Pulp", *Proceedings of the Symposium on Utilization of Hardwoods for Pulp and Paper*, FRI, Dehradun, pp. 103-112.

———. 1968b. "Production of Writing and Printing Papers from a Mixture of Central India Hardwoods", in *Proceedings of the Symposium on Utilization of Hardwoods for Pulp and Paper*, FRI, Dehradun, pp. 93-102.

Gupta, Tirath. 1978. "Nature of the Environment for Forestry Sector in India", *The Indian Forester*, 104(2) : 787-796.

———. 1979a. "Some Financial and Natural Resource Management Aspects of Commercial Cultivation of Irrigated Eucalyptus in Gujarat : India", *Indian Journal of Forestry*, 2(1) : 118-137.

———. 1979b. "Some Methodological and Practical Aspects of Accounting the Returns from Clearfelling the Natural Forests in the Evaluation of Intensive Forest Development Projects", *The Indian Forester*, 105(5) : 321-330.

———. 1984. "Demand-Supply Management and Pricing Policy for Cellulosic Raw Materials for Paper Industry in India", *Indian Journal of Agricultural Economics*, 38(3) : 291-299.

———. 1987a. "Choice of Tree and Grass Species for Social Forestry Plantations in Madhya Pradesh and Uttar Pradesh : Some Issues for Wasteland Management", in *Social Forestry Programme Management in India, Cases from Uttar Pradesh and Madhya Pradesh Volume I*, CMA, IIM. Ahmedabad, (Mimeo).

———. 1987b. "Protection of Social Forestry Plantations in Madhya Pradesh and Uttar Pradesh", in *Social Forestry Programme Management in India, Cases from Uttar Pradesh and Madhya Pradesh, Volume I*, CMA, IIM, Ahmedabad (Mimeo).

———. 1987c. "Management of Natural Vegetation on Degraded Lands: Some Lessons for Social Forestry Projects", in *Social Forestry Programme Management in India : Cases from Uttar Pradesh and Madhya Pradesh Volume II*, CMA, IIM, Ahmedabad (Mimeo).

Gupta, Tirath and Guleria, Amar. 1982. *Non-Wood Forest Products in India : Economic Potentials*, Oxford & IBH Publishing Co. Pvt. Ltd., New Delhi.

Gupta, Tirath and Mohan, Deepinder. 1982. *Economic of Trees versus Annual Crops on Marginal Agriculture Lands*, Oxford & IBH Publishing Co. Pvt. Ltd., New Delhi.

Gupta, Tirath and Sambrani, Shreekant, 1978. "Control of Shifting Cultivation : The Need for an Integrative Approach and Systematic Appraisal", *Indian Journal of Agricultural Economics*, 33(4) : 118-137.

Gupta, Tirath and Shah, Nitin. 1987. *Paper and Paperboards in India : Demand Forecasts and Policy Implications*, Oxford & IBH Publishing Co. Pvt. Ltd., New Delhi.

Guthrie, John A. 1973. *An Economic Analysis of the Pulp and Paper Industry*, Washington State University Press, Pullman.

Harrison, J.D.B. 1950. *Planning a National Forest Inventory*, FAO Rome.

ICICI (Industrial Credit and Investment Corporation of India). 1981. "Energy Utilization and Conservation in Industry", in *Workshop on Energy : Paper and Cement Industries—Summary Proceedings and Papers*, ICICI, Bombay.

IEMR. Year not known. *A Report on Requirement of White Printing Peper*, DCPPI, New Delhi.

IIPO (Indian Institute of Public Opinion). 1969. "The Growth Pattern of Paper Consumption : A Study for Self Reliance", *Quarterly Economic Report*, 16(2) : 10-19.

IPMA (Indian Paper Makers Association). 1979. *Directory of Indian Paper Makers Association*, IPMA, Calcutta.

——. 1982. *Directory of Indian Paper Makers Association*, IPMA, Calcutta.

——. 1983. *Workshop on Investments for Paper Industry : Proceedings*, IPMA, Calcutta.

IPPTA (Indian Pulp and Paper Technical Association). 1964. *Souvenir for International Seminar on Raw Materials for More Paper*, IPPTA, New Delhi.

Jacob, P.V. and Ramchandran, K. 1982, "Recycling of Waste : An INDAL Case Study", *Capital*, (Mid-Year Special) : 59-62.

Jain, Gyan Chandra. 1985a. "Indian Paper Industry-I : Performance of Various Units". *The Economic Times*, January 4.

——. 1985b. "Indian Paper Industry-II : Capacity Utilization Still Below Optimum Level", *The Economic Times*. January 5.

Jatia, J.M. 1982. *Presidential Address*, Annual General Meeting of the Indian Paper Mills Association, Calcutta.

——. 1983. *Presidential Address*. Annual General Meeting of the Indian Paper Mills Association, Calcutta.

Jha, S.S. 1981. "Cost Reduction in Paper Industry", *Management Accountant*, 16(8) : 366-69.

Joshi, N.J. 1982. "Forest Resources : Observations", *Proceedings : Seminar on Raw Materials for Paper Industry*, IPMA, Calcutta pp. 42-43.

Kapur, A.N. 1961. "Paper and Pulp Industries", *Journal of Industry and Trade*, 11(1) : 235-38.

Karamchandani, K.P. 1980. *Environment Management : Forestry*, Forest Department, Gujarat, Vadodara.

Keswani, S.L. 1982. "Cereal Straw as Raw Material for Paper Making in India", in *Papers Presented at Seminar on Raw Materials for the Paper Industry*, IPMA, Calcutta, pp. 149-159.

Khairoowala, Z. 1981. "Government Help Needed to Remove Paper Industry's Woe" *Industrial India*, 32(5) : 5-7.

Khetan, V.N. 1978. "Indian Paper Industry : Growth and Prospects", *Chemical Age of India*, 29(11) : 859-860.

———. 1979. "Shortage of Raw Materials", *Commerce*, 139(3569) : 931-933.

Kimmins, J.P. 1973. "The Renewability of Natural Resources", *Journal of Forestry*, 75(5) : 290-92.

———. 1976. "How to provide for Environmental Protection : By Regulation or by Use of the Price System", in *Timber Policy Issues in British Columbia*, edited by Mckillop, Williams. and Head, Water J., University of British Columbia Press, Vancouver.

Klemperer, W. David. 1976. "Economic Analysis Applied to Forestry : Does it Short-Change Future Generations?", *Journal of Forestry*, 74(9) : 609-611.

Kothari, P.S. 1971. "Royalty on Pulpwood", *The Indian Forester*, 97(6) : 293-297.

Krishnan, V. 1980. "Role of Mysore Paper Mills in Newsprint", *Commerce*, 141(3619) : 9.

Kundu, D.K. and Saha, N.K. 1981. "Approaches to Energy Conservation in Old Pulp and Paper Mills", in *Workshop on Energy : Paper and Cement Industries—Summary Proceedings and Papers*, ICICI, Bombay, pp, 93-111.

Mahalaha, S.H. 1971. "Natural Hardwoods for Pulping", *Conference on Utilizatian of Hardwoods for Pulp and Paper*, FRI, Dehradun. pp. 39-40.

Mahanti, P.C. 1978. "Paper : Rising Prices", *Commerce*, 137(3523) : 1012-1014.

———. 1984. "Paper Industry's Looming Crisis", *Commerce*, 148(1) : 312-315.

Maslekar, A.R. 1980. *A Study of Marketing of Wood Products in Maharashtra*, Indian Institute of Forest Management, Ahmedabad, (Mimeo).

Matharu, J.S. 1982a. "Forest Resources : Observations", in *Proceedings : Seminar on Raw Materials for Paper Industry*, IPMA, Calcutta, pp. 46-47.

———. 1982b. "Bagasse and Agricultural Residues : Keynote Address", in *Proceedings : Seminar on Raw Materials for Paper Industry*, IPMA, Calcutta, pp. 80-82.

Mathur, A.K. 1981. "Forest Resources : Observations", in *Proceedings: Seminar on Raw Materials for Paper Industry*, IPMA, Calcutta, pp. 48-50.

Mehta, M.M. 1981. *Towards Greater Energy Conservation and Optimum Capacity Utilization : Address at the 86th Annual General Meeting of IPMA*, IPMA, Calcutta.

Misra, D.N. and Suseta, Eladio. 1973. *Formulation and Economic Evaluation of a Bamboo Afforestation Project in South-East Uttar Pradesh*, The Ford Foundation, New Delhi (Mimeo).

Misra, D.N. 1980a. *Determination of Stumpage Price of Plantation Grown Crops: Case of Eucalyptus in Uttar Pradesh*, Programme Office, Indian Institute of Forest Management (IIFM), Ahmedabad (Mimeo).

———. 1980b. *Pricing of Eucalyptus Wood for Supply to Paper Mills : Process of Decision Making in Government*, Programme Office, IIFM, Ahmedabad, (Mimeo).

Misra N.D. 1973. "A Method of Grading Tropical Hardwoods", *Pulp and Paper International*, June, pp. 48-49.

———. 1978. "Hardwoods Pulping and Paper Making : Indian Experience", *Indian Pulp and Paper*, October-November.

Misra, N.D. and Raghuveer, S. 1985. "Energy Conservation, Rebuilding and Modernization in Paper Industry", *Commerce*, 50(3854) : 15-17.

Mukhedkar, P.A. 1982. "Paper Industry in Andhra Pradesh and Role of Forests", in *Papers Presented at the Seminar on Raw Materials for the Paper Industry*, IPMA, Calcutta, pp. 100-104.

Mukherjee, A.B. 1982. "Sunn Hemp as Raw Material for Making Speciality Papers", in *Papers Presented at the Seminar on Raw Materials for the Paper Industry*, IPMA, Calcutta, pp. 160-163.

Mukherji, A.K. 1982. "Forest Resources : Observations", in *Proceedings : Seminar on Raw Materials for Paper Industry*, IPMA, Calcutta, pp. 43-46.

Nair, V.S.K. 1982a. "Forestry Programme of Mysore Paper Mills Ltd.", in *Papers Presented at the Seminar on Raw Materials for the Paper Industry*, IPMA, Calcutta, pp. 122-126.

———. 1982b. "Forest Resources : Observations", in *Proceedings Seminar on Raw Materials for Paper Industry*, IPMA, Calcutta, pp. 51-52.

NAS (National Academy of Sciences). 1975. *Unexploited Tropical Plants with Promising Economic Value*, NAS, Washington, D.C.

———. 1979. *Tropical Legumes : Resources for the Future*, NAS, Washington, D.C.

———. 1980. *Firewoods Crops : Shrub and Tree Species for Energy Production*, NAS, Washington, D.C.

NCAER (National Council of Applied Economic Research), 1972. *Paper Industry ; Problems and Prospects*, NCAER, New Delhi.

NPC (The National Productivity Council). (No Date). *Report of the Study Team on Paper and Pulp Industry*, New Delhi.

Parkhe, M.S. 1981. "Introductory Observations" in *Workshop on Energy : Paper and Cement Industries—Summary Proceedings and Papers*, ICICI, Bombay, pp. 6-9.

Pearson, A.J. and Elder, R.A. *Preinvestment Study of Forest Resources : Mechanical and Chemi-mechanical Pulping of Eucalyptus*, UNDP-FAO, Rome.

Poddar, V. 1979. *Paper Industry in India : A Study*, Oxford & IBH Publishing Co. Pvt. Ltd., New Delhi.

Prasad, A.K. 1978. "R & D in Indian Paper Industry : A Critical Appraisal", *Chemical Age of India*, 29(11) : 889-96.

Raina, L.N. 1981. "Paper Industry : Victim of Neglect", *Eastern Economist*, 76 (4) : 255-264.

Raina, M.K. 1982a. *A Perspective for Paper*, Address at the 86th Annual General Meeting of IPMA held at New Delhi, IPMA, Calcutta.

———. 1982b. "Introduction", in *Papers Presented at the Seminar on Raw Materials for the Paper Industry*, IPMA, Calcutta, pp. 1-33.

———. 1982c. "Welcome Address", in *Proceedings : Seminar on Raw Materials for Paper Industry*, IPMA, Calcutta, pp. 1-3.

———. 1982d. "Forward", in *Proceedings : Seminar on Raw Materials for Paper Industry*, IPMA, Calcutta.

———, 1983. *The Paper Industry at Cross Roads*, Address at the 87th Annual General Meeting of the IPMA held at New Delhi, IPMA, Calcutta.

Rajan, T.P.S. 1985. "Problems of the Paper Industry", *Chemical Weekly*, 39(29) : 33-34.

———. 1982. "Paper in Plenty", *Eastern Economist*, 78(3) : 223-228.

Ranade, C.G. and Sah, D.C. 1983. *Industrial Utilization of Cotton Stock : Problems and Prospects*, Centre for Management in Agriculture (CMA), Indian Institute of Management (IIM), Ahmedabad, (Mimeo).

Rangan, S.G. 1982. "Raw Materials for Paper Industry", *Commerce*, 145(3725) : 7-11.

Rangarajan, C. 1978. "Determinants of Controlled Prices and their Implications", IIM, Ahmedabad (Mimeo).

Rao, A.N. 1971. "Paper, Pulp and Allied Industries", *Journal of Industry and Trade*, 21(8) : 8-15.

Ravindranathan, N. *et. al.* 1982. "Bagasse and Agricultural Residues for Paper Making in India", in *Papers Presented at the Seminar on Raw Materials for the Paper Industry*, IPMA, Calcutta, pp. 164-173.

Ray, P.K. 1982. "Raw Material Problems of Indian Paper Pulp", *Commerce*, 144(3707) : B 111.

Romm, Jeff. 1981. "The Uncultivated Half of India", *The Indian Forester*, 107(1 and 2) : 1-23, and 69-85.

Sadawarte, N.S. 1982. "Reclycling of Waste-Paper : A Pressing Need", in *Papers Presented at Seminar on Raw Materials for the Paper Industry*, IPMA, Calcutta, pp. 177-197.

Sagreiya, K.P. 1971. "Royalty on Pulpwood", *The Indian Forester*, 97(9) : 557-558.

Sarin, K.K. and Khanna, J.C. 1971. "Growth of Paper Industry in India", *Journal of Industry and Trade*, 21(8) : 31-35.

Sarkar, Subhash Chandra. 1975. "Paper Industry Poised for Rapid Growth", *Commerce*, 130(3344) : 1-9.

Saxena, S. 1982. "Paper : The Paranoia Persists", *Business India*, February 15-28, pp. 26-29.

Scolt, Alban E. 1980a. "Need for a Rational Policy", *Commerce*, 141 (3619) : 1-5.

———, 1980b. "Case for a National Policy on the Paper Industry", *Business Standard*, August 23.

Sen, D. 1978. "Prospects of Mini-Paper Projects in Indian Industries", *Chemical Age of India*, 29(11) : 862-864.

Seth, V.K. 1971. "The Promise and Problems of Indian Hardwoods as a Source of Raw Material for the Pulp and Paper Industry", *Conference on Utilization of Hardwoods for Pulp and Paper*, FRI, Dehradun, pp. 1-13.

———. 1982a, "Industrial Plantations : Observations: in *Proceedings : Seminar on Raw Materials for Paper Industry*, IPMA, Calcutta, pp. 58-59.

———. 1982b. "Role of Paper Industry in Meeting Its Raw Materials Requirements", in *Papers Presented at the Seminar on Raw Materials for the Paper Industry*, IPMA, Calcutta, pp. 84-94.

Seth, V.K. *et al.* (No Date). *Feasibility Study for Paper Mill in Baster Region of Madhya Pradesh*, Preinvestment Survey of Forest Resources Organization, Central Zone, Jagdalpur.

Sharma, L.C. 1978a. *Development of Forests and Forest Based Industries*, Bishan Singh Mahendrapal Singh, Dehradun.

———. 1978b. "Paper and Pulp Industry : An Economic Analysis", *The Economic Times*, November 24, Bombay.

SICOM (The State Industrial and Investment Corporation of Maharashtra). 1970. *Forest Industries : Scope for Development in Maharashtra*, SICOM, Bombay

Sinden, Jack A. and Worrell, Albert C. 1979. *Unpriced Values : Decisions without Market Price*, John Wiley, New York.

Singh, Hardev, 1980. "Paper Shortage Continues", *Eastern Economist*, 74(26) : 1278.

———. 1981. "No More Old Paper Machinery", *Eastern Economist*, 76(14) : 925.

Singh, M.N. and Mukherjee, V.N. 1965. "Fibrous Raw Materials for the Indian Pulp, Paperboard Industry", *Indian Forester*, 91(7) : 505-29.

Singhal, R.M. 1982a. "Forest Resources : Observations", in *Proceedings : Seminar on Raw Materials for Paper Industry*, IPMA, Calcutta, pp. 36-39.

———. 1982b. "Forest Raw Materials for Paper Industry in Madhya Pradesh", in *Papers Presented at the Seminar on Raw Materials for the Paper Industry*, IPMA, Calcutta, pp. 113-121.

Singhania, Hari Shankar. 1982. "Introductory Observations" in *Proceedings : Seminar on Raw Materials for Paper Industry*, IPMA, Calcutta, pp. 4-5.

Sinha, G.M. 1971a. "How to Ensure Enough Raw Materials for Paper Units", *The Financial Express*, December 7.

———. 1971b. "Paper and Pulp Industry", *Journal of Industry and Trade*, 21(8) : 42-46.

Sinha, N. 1980. "Technological Inputs for Fuller Capacity Utilization", *Business Standard*, August 23.

Solow, Robert M. 1974. The Economics of Resources or the Resources of Economics", *The American Economic Review*, 64(2) : 1-14.

Somani, B.D. 1971a. "Comparative Study of Pulp and Paper Industry in Japan and India", *The Financial Express*, March 22.

———. 1971b. "Paper and Pulp Industry", *Journal of Industry and Trade*, 21(8) : 42-46.

Somani, K.K. 1982a. "Bagasse and Agricultural Residues : Observations", in *Proceedings : Seminar on Raw Materials for Paper Industry*, IPMA, Calcutta, pp. 89-90.

———. 1982b. Social Forestry for the Paper Industry", in *Papers Presented at the Seminar on Raw Materials for the Paper Industry*, IPMA, Calcutta, pp. 95-99.

Somani, Ravindra B. 1975. Difficulties posed by Production Control Order", *Commerce*, 130(3344) : 11.

———. 1980. "Ban Import of Coated Paper", *Commerce*, 141(3619) : 8.

Sowani, M.Y. 1982a. Forest Raw Materials for Paper Industry in India", in *Papers Presented at the Seminar on Raw Materials for the Paper Industry*, IPMA, Calcutta, pp. 37-50.

———. 1982b. Industrial Plantations : Keynote Address", in *Proceedings : Seminar on Raw Materials for Paper Industry*, IPMA, Calcutta, pp. 56-57.

Srivastava. T.N. and Guha, S.R.D. 1968. "Indian Pulp and Paper Industry", *Proceedings : of the Symposium on Utilization of Hardwoods for Pulp and Paper*, FRI, Dehradun, pp. 62-68.

Strong, M. 1973. "One Year After Stockholm : An Ecological Approach to Management", *Foreign Affairs*, 51 : 690-707.

———. 1974. "The Case for Optimum", *Saturday Review*, 14 : 7-11.

Sundaram, Satya 1980. "Paper Industry : Problems and Prospects—I", *National Investment and Finance Weekly*, 38(4) : 205-210.

Swaminathan, M.S. 1982. "Forest Resources : Key Note Address", in *Proceedings : Seminar on Raw Materials for Paper Industry*, IPMA, Calcutta, pp. 14-20.

Tapadar, D.C. 1982. "Bagasse and Agricultural Residues : Observations", in *Proceedings : Seminar on Raw Materials for Paper Industry*, IPMA, Calcutta, pp. 83-87.

Thapar, L.M. 1982. "Forest Resources : Keynote Address", in "*Proceedings: Seminar on Raw Materials for Paper Industry*, IPMA. Calcutta, pp. 20-26.

Thapar, S.D. 1974. "Second India Studies : Forestry in Year 2001", The Ford Foundation, New Delhi.

———. 1975a. *India in 2000 : Second India and Forestry*, Association of Voluntary Agencies for Rural Development, New Delhi.

———. 1975b. "Forestry in 2001", *Voluntary Action*. 17(2) : 34-35.

Tiwari, K.M. 1982a. "Raw Materials for the Paper Industry", in

Papers Presented at the Seminar on Raw Materials for the Paper Industry, IPMA, Calcutta. pp. 51-57.

——. 1982b. "Bagasse and Agricultural Residues : Observations", in *Proceedings : Seminar on Raw Materials for the Paper Industry*, IPMA, Calcutta, pp. 51-57.

Tiwari, N.D. 1982. "Inaugural Address", in *Proceedings : Seminar on Raw Materials for Paper Industry*, IPMA, Calcutta, pp. 7-13.

Vandana, Shiva. *et al.* 1982. "Social Forestry : No Solution with the Market", *Ecologist*, 12(4)

Vermah, J.C. and Pant, M.M. 1981. Production and Utilization of Bamboos" *The Indian Forester*, 107(8) : 465-476.

Vohra, B.B. 1975. *Land and Water Management Problems in India*, Training Division, Department of Personnel and Administrative Reforms, GOI, New Delhi.

Waheedkhan, M.A. 1971. "Determination of Royalty, on Pulpwood", *The Indian Forester*, 97(9) : 558-561.

Webster, H.H. and Gordon, J.C. 1975. "Timber, Time and Money : An Opinion", *Journal of Forestry*, 73 : 265-267.

Westoby, Jack C. 1963. *The Role of Forest Industries in the Attack on Economic Under Development*, FAO, Rome (Mimeo).